Mesozoic Fossils I
Triassic & Jurassic Periods

Bruce L. Stinchcomb

Schiffer Publishing Ltd®

4880 Lower Valley Road, Atglen, Pennsylvania 19310

Dedication

This book is dedicated to my parents, Leonard and Virginia Stinchcomb, who first introduced the author, as an adolescent, to Mesozoic rocks and fossils of the Cretaceous Owl Creek Formation.

Other Schiffer Books by Bruce L. Stinchcomb
Paleozoic Fossils. ISBN: 9780764329173. $29.95
World's Oldest Fossils. ISBN: 9780764326974. $29.95

Other Schiffer Books on Related Subjects
Whales and Seals: Biology and Ecology. Pierre-Henry Fontaine. ISBN: 9780764327919. $34.95

Copyright © 2009 by Bruce L. Stinchcomb
Library of Congress Control Number: 2008936057

Designed by Mark David Bowyer
Type set in Bodoni Bd BT / Zurich BT

ISBN: 978-0-7643-3163-3
Printed in China

Schiffer Books are available at special discounts for bulk purchases for sales promotions or premiums. Special editions, including personalized covers, corporate imprints, and excerpts can be created in large quantities for special needs. For more information contact the publisher:

Published by Schiffer Publishing Ltd.
4880 Lower Valley Road
Atglen, PA 19310
Phone: (610) 593-1777; Fax: (610) 593-2002
E-mail: Info@schifferbooks.com

For the largest selection of fine reference books on this and related subjects, please visit our web site at
www.schifferbooks.com
We are always looking for people to write books on new and related subjects. If you have an idea for a book please contact us at the above address.

This book may be purchased from the publisher.
Include $5.00 for shipping.
Please try your bookstore first.
You may write for a free catalog.

In Europe, Schiffer books are distributed by
Bushwood Books
6 Marksbury Ave.
Kew Gardens
Surrey TW9 4JF England
Phone: 44 (0) 20 8392-8585; Fax: 44 (0) 20 8392-9876
E-mail: info@bushwoodbooks.co.uk
Website: www.bushwoodbooks.co.uk
Free postage in the U.K., Europe; air mail at cost.

Contents

Acknowledgments

The author wishes to acknowledge and thank the following persons for various forms of assistance, specifically assistance in various ways regarding Mesozoic fossils and strata in the genesis of this book: Matthew Forir, Allan Graffham, Richard and Lisa Hagar, Steve Jones, Curvin Metzler, John Stade, and Dennis Whitney. For original artwork, my daughter Elizabeth Stinchcomb and for graphics Patricia Eicks is acknowledged.

Introduction

This is the author's third book on fossils with an emphasis toward collecting these "medals of creation." My previous works on this subject, "Worlds Oldest Fossils" and "Paleozoic Fossils," dwelt with earlier parts of geologic time. This book is the first of two works dealing with fossils of the Mesozoic Era; it deals specifically with those fossils of the Triassic and Jurassic periods. Besides presenting a plethora of Triassic and Jurassic fossils from around the world, the book takes a somewhat historic approach. This approach is taken in light of the fact that the understanding of fossils and strata and the fundamental part that they play in the understanding of the earth's history took place in the Jurassic rocks of England, France, and Germany. A chapter on large Mesozoic vertebrates and world's fairs compliments this historic theme.

Chapter One

Mesozoic Fossils, Fossil Collecting, and Strata

The "Jurassic" Coast of England and the Jura Mountains of Germany

Mesozoic is the "Middle Era" of geologic time and, like the other parts of the (Phanerozoic) geologic time scale, was based upon fossils and related strata explored during the 1830s and '40s when the sequences of rock strata in England were first being worked out. The Mesozoic Era became known as the "Middle Era," that part of geologic time known also as the "age of reptiles." The **eras** of geologic time being worked out (and their subdivisions, the geologic **periods**) were established on the basis of fossils present in rock strata under investigation. It's important to realize that the persons doing this documentation viewed the underlying constructs regarding the succession of life as documented by fossils differently. It was different from geologists of the twentieth century and today as these early geologists were creationists, creationists who believed that the world represented by the fossils being found were from a "pre-adamite" world in which saurians, like dinosaurs and ichyosaurs, represented Gods highest creations in this part of the earth's past. To the geologists working on strata of this reptilian "middle era," it became obvious that rock strata occurring along the south coast of England rested upon older, more disturbed strata found farther to the north, strata that contained the seemingly more primitive fossils of the Paleozoic Era. Grey beds of shale, slabby limestone, and sandstone, making up the sea cliffs of southern England contained an abundance of shelly fossils as well as the much rarer saurians. The abundant fossils, known as ammonites, had been widely collected throughout the eighteenth century in areas such as Lyme Regis to the south of London and Whitby to the north. These coastal towns were also favorite places for an English "holiday," being at most a few days coach ride from London, where one could enjoy the warm, salty waters on a summer's day. The collecting and sale of fossils and sea shells ("She sells sea shells by the sea shore"), particularly catering to the seaside tourist crowd, developed into a busy cottage industry in the late eighteenth century.

In the 1820s, near the town of Lyme Regis, after some particularly fierce winter storms knocked down large parts of the sea cliffs, a fossil and sea shell monger by the name of Mary Anning found, in the newly fallen slabs, a complete lizard-like animal and later another even larger fossil saurian, both of which she sold for a "pretty penny" to geological tourists. These two fossil "saurians" were similar to fossils being found in southern Germany, in the Swabish "Albs" or Jura Mountains of Bavaria and Wurttemburg. These German beds contained, as did those of England, numerous ammonites, a fossil whose origin had been debated extensively in the eighteenth century, and whose ending with "ite" indicated that it was found in and came from the Earth.

The saurians collected by Mary Anning found their way to geologic pioneers such as William Buckland, Adam Sedgwick, and Charles Lyell. The latter would author a book in 1832, *Principles of Geology*, which would have wide distribution and influence. This book, if you disregard the earlier "hard to read" manuscript of James Hutton, is considered as the earliest book outlining modern geology with its emphasis on fossils and strata, a prerequisite to that abstract concept so important in modern natural history known as "geologic time."

Lyell outlined these "middle era beds," which included, beside ammonites, discussions on the fossil saurians found in the Lyme Regis cliffs as well as similar ones found in strata informally called "Jurassic" after Germany's Jura Mountains. In the 1840s, additional saurians came to light, which included the "terrible lizards" or dinosaurs of Sir Richard Owen and Dr. Gideon Mantell. On the other side of the Atlantic similar fossils were found near Haddonfield, New Jersey, by Joseph Leidy so that in 1841, on a trip to North America, Charles Lyell would become acquainted first-hand with these saurians and the strata with yielded them. A second trip to North America, seven years later, in 1847-48 included Lyell's examination of saurian bearing strata in Georgia and Alabama in bluffs along the Alabama and Tombigbee Rivers.

As an accompaniment to a somewhat historical approach taken in this book, many woodcut illustrations from Lyell's geologic works are reproduced, often next to similar fossils which came from the same outcrops and strata as mentioned by Lyell. Many of the saurians found in Britain and Germany were from the "middle era" or Mesozoic strata of geologic time and included the Ichyosaur, the Plesiosaur, and the dinosaur, as well as the Mosasaur, specimens of which had been discovered in the eighteenth century in a chalk quarry near the town of Maastericht in the Netherlands. All of these large saurians disappeared abruptly from rock strata deposited at the same time as did the more abundant ammonites and other mollusks like *Inoceramus* clams and belemnites. This sudden termination of Mesozoic fossils, all at a distinct point in time, defined the "top" (related to the top of strata) or the end (in time) of what would become known as the Mesozoic Era.

Strata like that at Lyme Regis along the southern coast of England occurred beneath strata that yielded the mosasaurs of the Netherlands and because it was beneath this mosasaur bearing strata was of course older than the strata yielding the mosasaurs. Strata with the same fossils as at Lyme Regis were found to be even better represented in the Jura Mountains of Bavaria and Baden-Wurttemberg in southern Germany and were named Jurassic after the Jura Mountains of that region. Strata representative of what would become the "bottom" or the beginning of the Mesozoic Era was also best represented in Germany where such strata was found to contain fossils like ceratitic ammonoids and saurians different from those found in strata being called Jurassic.

Such strata came in three distinct parcels, a lower reddish layer (Bunter), an overlying fossiliferous limestone (Muschelkalk) and younger strata than the Muschelkalk—a saurian bearing sequence known as the Keuper. Such a mega-sequence was named Triassic after these three (tri) horizons. Underneath the Triassic beds occurred strata which included the obviously older and more deformed layers (that is rock layers that tilted, folded, or crumpled), as seen in Wales, Bohemia, and Poland, and given the name of Paleozoic (Ancient life).

With the Mesozoic Era one takes a trip into the realm of dinosaurs and pterodactyls as well as other extinct saurians like the Mosasaurs. In terms of fossil collecting, fossils of these animals are in quite a different category than most others. For starters, such fossils can be quite large and bulky, as many of the animals of which they were once a part were quite **large**. Potentially here also can be the "big money" as the public has a fascination with these fossils to an extent greater than it does with most other fossils. Such popularity, and its accompanying economics, is a combination that can attract the bad aspects, which other occurrences of "big money" can also attract. Probably nowhere else (other than in hominid fossils) is this monetary aspect as prevalent as it is with Mesozoic vertebrates. As with other sides of paleontologic collecting, there are both good and bad aspects in this for the serious and sincere collector.

The Triassic Period was named for the three distinct divisions of its strata found in Germany. These are the Bunter, a red sandstone, the Muschelkalk, and the Keuper, a series of olive drab, marine shales. This outcrop of red sandstone is similar to the Bunter of Germany. Red sandstone and siltstone beds are characteristic of Triassic strata worldwide. Such "red beds" generally have few fossils, however limestone beds and grey or olive drab shale or siltstone beds deposited in a marine environment can be a source of the relatively hard-to-get fossils of the Triassic Period. These outcrops are in Nevada.

Spreading the Paleontological Record

Fossils, sometimes quite desirable ones, occur in beds or layers in considerable abundance. Such "fossil beds" may (or may not) become known and may be looked upon as a local or even as a national "treasure." At other times a unique fossil resource can be virtually ignored and considered inconsequential. Reasons for the focus on particular beds vary, including ease of access, recognition by media, inclusion of popular fossil species, and the actions of motivated individuals, among others. A common development with significant fossil resources is to set up a local museum and to see that specimens are housed in it. Following such "museum awareness" may occur the mindset of "this is a local treasure so we cannot allow our fossils to leave the local area (or the county of origin)."

The Painted Desert, part of the Petrified Forest National Monument. The Painted Desert consists of badlands of the Upper Triassic Chinle Formation, a series of red shales, mudstones, and sandstone eroded into a series of gullies by running water in an environment with low levels of precipitation. The badlands are predominantly red in color, a characteristic of Triassic rocks over most of the world.

This protectionist attitude regarding fossil resources has become more prevalent worldwide during the last few decades. This author suggests a more useful strategy, rather than such a protectionists one, would be to work the fossiliferous layer (or layers) cooperatively and exchange specimens thus acquired for those from other parts of the world. Another strategy might be to let the "magic" of the marketplace work, selling the common fossils ac-

quired. In this case, the pursuit of monetary gains will come into play, with all that this entails, but this could be useful, if so desired, in the acquisition of specimens from other parts of the world. Working with an abundance of fossils is an excellent method of becoming knowledgeable within the field of paleontology—especially knowledgeable in the way that comes through permanent learning, which is associated with such "hands on activity." Exchanges of specimens, rather than protecting them by all "locking them up" either in the ground or by placing them in endless repetition in museum drawers, could represent a type of paleontological "cross fertilization" between professionals and amateurs that would ultimately produce the greatest benefits for all, and that would represent a classic example of a win-win situation for everyone.

In the author's admittedly somewhat biased opinion, when paleontologists employ this protectionist worldview, what is at times an abundant (if localized) resource is removed from public access. This creates a great deal of tension between the paleontologists and the collecting community, a tension that in my opinion cannot be good for the future of paleontology. Further, this perspective *limits* the flow of information concerning fossil discoveries around the globe. While the desire to preserve resources such as fossils is commendable, I believe it is often biased by that archaeological model used by some paleontologists, who come to view fossils as extremely limited resources. I contend that many desirable and significant fossils occur worldwide. Greater access and exposure to a broad range of them (as can be done vicariously in this book) is of great educational benefit, and education is a fundamental reason for collecting. I would like to call for greater cooperation between these two camps as there are too few professionally trained paleontologists to recover all the fossils endangered by the effects of erosion, construction, and time, certainly the two groups would gain more through cooperation than controversy.

Extent of the
Mesozoic Fossil Record

The fossil record is vast and increases in diversity the closer one gets to that part of geologic time approaching the present. The goal of this book (as well as previous and future books by the author) is to present a first-order view of portions of the fossil record with the collector in mind. Few microfossils or other of the more technical elements, which can constitute a major part of the fossil record and paleontology, are represented. There is a bias toward fossils that are obvious, attractive, and collectable — science needs its purveyors of that which is attractive, and fossils can be things of beauty that can delight persons who otherwise might be cool to their scientific aspects.

Diversity of fossils, such as those found in Mesozoic age rocks, presents the problem of what should be included in a diverse work such as this, as is the maintenance of stratigraphic and taxonomic accuracy. The author's expertise in taxonomic detail resides with Cambrian and Lower Ordovician fossils. He is knowledgeable in Mesozoic and other younger records, but not to the depth that he has with some earlier parts of geologic time. Identification as to genus and species (when present) was often done by reliance on either general works on paleontology such as *Index Fossils of North America*, or the *Treatise of Invertebrate Paleontology*. Identification of some taxa at times was done by using identifications accompanying the fossil specimens themselves; specimens obtained either through trade or purchase. For those concerned with identifying a fossil "exactly," it should be mentioned that in modern paleontology the trend has been more and more toward a fine tuning of the taxonomic groups represented, with paleontologic "splitters" often being prevalent. Such splitting is justified if it can emphasize slight evolutionary changes, or if it can delineate paleontologic zones and other related biologic aspects of stratigraphy.

At other times, such splitting is less scientifically justifiable and serves only to confuse persons delving into the literature of paleontology; with this in mind, the author has not always included the work of the "splitters."

Redbeds of the Triassic Karoo Series of South Africa. The Karoo Series is a sequence of red beds that, in some areas of outcrop, contain numerous strange Triassic fossils. This includes the most complete record of the Theraspids or mammal-like reptiles, a group of reptiles that gave rise in the late Triassic to the mammals. Therapsid fossil bones can be locally common in some beds; however, it is unfortunate that the South African government prohibits collecting, ownership or export of **all** fossils.

Petrified logs near Winslow, Arizona. The Chinle Formation can locally contain large numbers of petrified logs which weather from the red sediments of badlands of the Chinle Formation, particularly a Middle Triassic gravel bearing member of the Chinle called the Shinarump Conglomerate, which forms the base of the Chinle Formation. Most of these logs were from conifers such as sequoia. They are not found erect but rather were washed into the sediments by floods and are essentially fossil driftwood. This petrified conifer wood is found in considerable quantities in shale and conglomerate (cemented gravel) beds of northern Arizona and southern Utah. The best-known area where these logs occur is in the Petrified Forest of northern Arizona where these logs are located.

This is an outcrop of the middle part of the Triassic strata of Germany, the Muschelkalk Formation. It can be the source of fine fossils in Germany such as crinoids, ceratitic ammonites, and unusual marine vertebrates (Placodonts).

The Keuper Formation forms the top or youngest beds of Triassic strata in Germany. It consists of olive drab, marine shale and siltstone beds like that shown here. The Keuper has produced a number of large ammonites and some unusual marine reptiles.

These are Lower Jurassic age, massive sandstones (Navajo Formation), formed under arid conditions in what is now the American Southwest. Such sandstone beds rarely contain fossils; those few that do occur are trace fossils (tracks or trackways), which occur on infrequent bedding planes.

Tilted, massive beds of Lower Jurassic Navajo Sandstone at Dinosaur National Monument, Vernal, Utah.

Lower Jurassic, Navajo Sandstone. This desert sand dune deposit consists of massive layers of cross-bedded sandstone. Fossils are absent in these sandstone beds, except for trackways of dinosaurs found on the infrequent bedding planes.

Middle Jurassic gypsum beds of the Sundance Formation, western Black Hills, South Dakota. Gypsum beds like this were deposited in a hypersaline environment, an environment in which few organisms could live. Jurassic marine strata becomes more and more representative of hypersaline environments in the eastern areas of outcrop in western North America. The outcrops shown in the previous photo (Middle Jurassic shale, siltstone, and sandstone beds) represent a more normal marine environment and a type of rock that can contain marine fossils like belemnites. Hypersaline deposited strata like the gypsum beds shown here lack any obvious fossils. The black object on the outcrop is a light meter measuring about three inches long.

Middle Jurassic shale, siltstone, and sandstone beds of the Sundance Formation, southern Idaho.

In many instances older names of fossils are the better known ones and sometimes these become well established. In this work, the author has usually used such established names, in part because he is not familiar with the (often) vast amount of more recent literature done by splitters. This is done also as the taxonomy presented by splitters can be complex and specialized and may be subject to change as further evidence is recovered from the fossil record. What sometimes happens is that an author will introduce a new name for some well established group, then another author will come along and invalidate that which was previously done, that is, he puts the "new" taxa in synonymy, in some cases going back to the original name. One of the groups where this type of taxonomy is prevalent is in the ammonites, those mainstays of Mesozoic strata.

Ammonite diversity is potentially immense, with slight variations in ornamentation, suture patterns, and other variables used in varying degrees in ammonite taxonomy. As the ammonites were a successful group of organisms, — one that existed for tens of millions of years, a large number of forms (and hence names) is perhaps not unreasonable. The mastery of such a diversity of species might require many years of study for just one particular group. As a consequence of this type of phenomena, modern paleontology has sometimes become highly specialized. As there are many groups of fossils represented by this model besides ammonites, total knowledge of them all is nearly impossible. This phenomena is the reason why most paleontology today requires such a high level of specialization, a specialization so great that a generalist may sometimes no longer feels welcome.

Diversity in Ammonites, a Particularly Noteworthy Mesozoic Group of Mollusks

This diversity in some fossils applies to the identification of ammonites worldwide, including the rich ammonite faunas of Madagascar, a large number of which have recently come onto the fossil market. This fauna has not been studied to the same extent as have the British or German ammonite faunas. Superficially, however, the Madagascar and European faunas appear to be rather similar; however, an in depth study of the two faunas, when done, might find significant differences between them. The author has taken the approach of using European ammonite names and concepts to this ammonite fauna and others found on the other side of the world, a strategy that may or may not be appropriate but is reasonable as ammonites, or their empty shells, could travel long distances over past oceans.

Middle Jurassic cross-bedded dune sandstone, Madagascar. Mesozoic strata in some parts of this large island can be somewhat similar to those of Europe or, in this case, similar to Lower Jurassic strata of the western United States (Navajo Sandstone).

Close-up of Middle Jurassic cross-bedded dune deposited sandstone, Madagascar.

Pros and Cons of Collecting Vertebrate Fossils, Particularly Large Ones

Collecting Mesozoic fossils brings up the issue of the appropriateness of personal collection of fossil vertebrates, particularly the collection of large vertebrate fossils. As with most controversial issues, there are multiple positions, which can range at one end from "absolutely not under any circumstance" to that of the private collector who has "squirreled away" unique and complete specimens, which have not, but should be, scientifically investigated. Some museums and their personnel take the "absolutely not" position, as do some key members of SVP (Society of Vertebrate Paleontology). Many geologists, on the other hand (which includes the author), take a more liberal position on this matter. Besides considerations regarding the bulk and weight of vertebrate fossils, especially large ones, collecting fossil vertebrates can be a different matter

from collecting normally small fossils such as plants and invertebrates. Large vertebrates are almost always found in a fragmented condition, consisting of individual pieces. Such individual pieces can be compared with the parts of a jigsaw puzzle, and, as with a puzzle, the more pieces you have, the easier it will be to assemble the puzzle and the more accurate will be your results.

Large vertebrates hold a fascination for the public which is usually greater than that of more "normal" fossils. The author periodically gives "show and tell" talks to schools and can attest to the enthusiasm that a dinosaur bone, a tooth or an egg can incite with a large part of the public, particularly with children. The position that no such activity should take place outside the jurisdiction of institutions and that non-institutional engagement with fossil vertebrates is entirely inappropriate is one the author considers extreme and negates educational aspects that vertebrate fossils can offer.

As to the matter of the appropriateness of a private cabinet which includes vertebrates (or parts of such), the matter has to be taken up on a case by case basis. Not all large vertebrate fossils are rare, and this includes many dinosaur fossils. The matter of collecting them also has to consider the collector's role in the preservation of such fossils, which might otherwise be destroyed. Such a case-by-case appropriateness in the matter of vertebrates is addressed in this book under the particular fossil being discussed.

Archeology, Dinosaurs, and Mesozoic Fossils

With the matter of dinosaurs as well as other Mesozoic vertebrates, it has recently become somewhat common to find references to archeology and dinosaurs. Archeology deals with the unwritten record of man, a record covering a time span of at most five million years. Dinosaurs went extinct some sixty-seven million years ago, so it should be obvious that no "man" has ever seen or had anything to do with a living dinosaur. This it seems, however, doesn't stop the association of archeology with dinosaurs.

Part of this dinosaur-man association appears to be intentional disinformation and comes from two sources. One of these is the Young Earth Creationists who claim that man and dinosaurs were contemporaneous and that the Earth is only a few thousands of years old. As an example of one of their "proofs" of this they cite a widely circulated urban myth regarding the occurrence of a gold chain found in a Mesozoic coal seam. Even more insidious are publications, such as John D. Morris's *Tracking Incredible Dinosaurs and the People Who Knew Them*.

Upper Jurassic Solnhofen Plattenkalk, a thinly layered, marine limestone deposited in a lagoon located behind Jurassic coral reefs. This limestone is worked in many quarries in southern Germany where it is used as decorative rock in building interiors as well as used as the stone which forms the basis of lithography (literally writing on stone). The Solnhofen Plattinkalk has produced a considerable variety and number of fine, well preserved fossils, including the toothed bird *Archeopterix lithographica*, considered by some to be the worlds most valuable fossil. the Solnhofen Plattenkalk is one of the paleontologic "windows" of geologic time.

Upper Jurassic, non-marine deposits, Morrison Formation. This series of stream deposits in the western U.S. is famous for its late Jurassic dinosaurs as well as other fossils such as plants. The Morrison Formation is also the source of many of the uranium minerals mined on the Colorado Plateau. This tilted outcrop is in Dinosaur National Monument near Vernal, Utah. In the background can be seen large hills made up of the underlying and therefore older (Lower Jurassic) Entrada or Navajo Sandstone.

Outcrop of the Late Jurassic Morrison Formation containing many exposed dinosaur bones. The outcrop is protected by a large building built over the tilted mudstones of the Morrison Formation at Dinosaur National Monument, Vernal, Utah.

The other connection with dinosaurs and pre-historic man appears to come from persons who do not understand megatime and confuse the time scales of archeology and geology. As stated before, archeology deals with man, and **man** (in one form or another) has been around for **at most** five million years and dinosaurs went extinct at the end of the Mesozoic Era, sixty-seven million years ago. Life has existed on the Earth for some four billion years and this humongous time span can be the realm of paleontology but is **not** that encompassed by archeology. The author has observed, particularly since the popularity of dinosaurs and *Jurassic Park*, numerous references to archeologists dealing with dinosaurs in popular literature. Confusion of archeology with paleontology (and therefore artifacts with fossils) is common and has, at times, been the reason for restrictive legislation and policies regarding fossils. Many countries which have rich archeological and ancient historical resources, which they understandably wish to protect, also

have strict prohibitions regarding the collection of, export, and even ownership of fossils; a legislative byproduct of not understanding the distinction between fossils and artifacts. Professional archeologists are well aware of the difference between fossils and artifacts, however the average person usually isn't—the archeological model applied to fossils seems logical to large numbers of the public—the thinking being that something abundant that is millions, or hundreds of millions of years old should be given the same legislative protection as something found in small numbers that is thousands or tens of thousands of years old.

It should be mentioned that the current US federal policy (as of 2008) regarding paleontological and geological collecting, allows for casual, non-commercial surface collecting of non-vertebrate fossils on National Forest and BLM (Bureau of Land Management) land without a permit. In many states the policy can be quite different, however, where the archeological model of site management and collecting is applied to fossils and permits for collecting are required, which may be difficult to obtain.

On the matter of Mesozoic Era fossils (and younger ones as well), large vertebrates like dinosaurs, if found on public land, should generally be left alone and/or reported to a suitable person such as a museum representative or an official with a state or provincial geological survey. These larger vertebrates generally require specialized knowledge for their collection and they are the ones to which paleontological collection restrictions are often aimed at.

At this point it might be noted that fossils, sometimes nice ones, can turn up in unexpected places, particularly after one develops an eye for them. Not only rock outcroppings along streams, but sometimes rock piles and even gravel beds can be a source for interesting specimens.

Participatory Paleontology vs. Spectator Paleontology!

Collecting fossils offers a "hands on" method of involvement in paleontology! Often persons get interested and involved with paleontology through the finding and collecting of fossils. Such direct hands-on involvement has the effect of implementation of learning of a permanent type, which can also involve a permanent interest in paleontology. With Mesozoic fossils one also is immersed into the world of dinosaurs as well as other large vertebrates, which are given the popular name of "ruling reptiles." Such fossils can hold considerable interest for a large portion of the public, especially since the *Jurassic Park* movies and books. Natural history museums have traditionally been the vehicles for presentation of such large fossils to the public; however, the entertainment industry has also gotten into the act during the past few decades. With both of these venues, the viewpoint generally taken is that of the public being a spectator. Some persons, however, want to get involved in a way that is more personal and to participate directly in some of the action. Collection and interaction in paleontology with fossils is a logical and effective way to do this. Fossils can be locally common and fossil collecting is a participatory, "hands on" type of activity, and is a great way to get to know about past life in a setting involving permanent learning. Dinosaurs and other large vertebrates, however, are some of the least suitable fossils for an individual to collect; they are bulky and individual bones are not particularly attractive. Also, individual parts of a large vertebrate may

Prior to becoming a National Monument these exposures of volcanic tuff were made available for fossil collecting to persons taking in what was then the privately owned Colorado Petrified Forest. This hillside outcrop was a great learning tool for persons of all ages where they could easily collect nice fossils. An abundance of fossil insects was one of the rewards of collecting as well as fossil angiosperm leaves. Such hands on activity, as seen here, ceased when the area became Florissant Fossil Beds National Monument to preserve the site from being sold for land development. The U.S. Park Service has a strict policy of no collecting of any kind by the general public. Such a protectionist policy, while saving the area from development also placed "off limits" a very effective "hands on" educational tool. Such a view toward paleontological sites uses an archeological-site protection model applied to paleontology which, in most instances, is inappropriate. Inappropriate because fossils occur in beds of strata which usually extend over a considerable area. For instance the fossil bearing, Florissant tuffs shown here, extends under the soil over many square miles, the hillside shown here is just where these fossil bearing beds have been exposed. The site has now been soil covered and grassed over.

represent only a portion of a skeleton and collecting only parts of it can be the equivalent of losing some significant parts of a jigsaw puzzle. This problem has been stated before and offers a reasonable cause for vertebrate paleontologists to discourage individual collecting with its participatory advantages. Smaller vertebrate fossils, however, do not carry such baggage and for a number of reasons are more suitable for individual collecting. Here also, however, can be found opposition to individual involvement in collecting that, unlike resistance associated with large vertebrates, is usually not as justifiable.

The matter of the personal collection of fossils involves many issues, some which can really be a "can of worms." Suffice it to say that one should use common sense, try to obtain landowner and manager permission to collect, and become acquainted with other persons with an interest in fossils. Often rockhound groups or amateur geology societies occur in a region and involvement with such groups can be a great way not only to know the "ropes" on local collecting but also to make new friends with a common interest.

Besides personal collecting, various mineral and fossil fairs offer a type of "hands on" opportunity for an individual to become involved with fossils. It is with some sadness that one encounters some institutional representatives who regard such fairs with contempt, considering that the persons involved in them generally love what they are doing and that the fossils and minerals presented are available to all, institutional and private collector alike.

More on Mesozoic Fossils and Collecting

Mesozoic strata can hold a variety of fossils! Invertebrates such as mollusks, echinoderms, and corals, as well as other marine life can be common in Mesozoic marine strata and fossil mollusks can even be common in non-marine layers. What often differentiates Mesozoic rocks from earlier ones is the presence of vertebrate remains, particularly the teeth and spines of sharks in marine strata, and the bones and teeth of dinosaurs and other ruling reptiles in non-marine layers. The presence of fossil vertebrates in abundance is one characteristic of Mesozoic strata and this may set it apart from earlier strata. So-called bone beds do occur in earlier strata, however it is in the Mesozoic where these become more obvious and tempt the collector.

A fossil site! Fossil collecting offers a particularly effective vehicle for persons to become interested in science. Termination of participatory "hands on" activity as seen here and replacing it with a policy where members of the interested public are involved with fossils as spectators only is becoming the norm. This "spectator only" model has been applied to many famous fossil localities in the U.S. and Canada, where the archeological model of preservation has been applied. Creek outcrops like this can be great places for fossils and not all are off limits.

Vertebrates as
Multi-component Fossils

This presence of vertebrate fossils in Mesozoic rocks presents a dilemma with the collector and collecting. Vertebrates, being multi-component fossils, can have a complete skeleton in an area where only a few bones might give its location. Complete vertebrate skeletons are scientifically valuable and rare fossils; a knowledgeable collector abides by the dictum that "scientific consideration" comes first. On the other hand, contrary to comments by some vertebrate paleontologists, individual bones and teeth of many vertebrates can be quite common locally and not necessarily of any scientific value. The matter of a few fossil bones being a clue to the presence of a complete vertebrate skeleton brings up some thorny issues regarding collecting but the archeological model of "no non-institutional collecting" of vertebrate fossils often is an inappropriate one. Inclusion of vertebrate fossils in this work, a compendium for fossil collectors and fossil collecting, was approached with trepidation for this reason, but the realization that vertebrates form a significant portion of the Mesozoic fossil record and that interest in them is high necessitated their inclusion. The common occurrence of many vertebrate fossils like shark and ray teeth, small fossil fish, and even partial dinosaur bones makes opposition to the collecting of **all** vertebrate fossils somewhat irrational. The author's exposure to Mesozoic strata and its associated fossils was enhanced by his involvement in the commercial geologic exploration for uranium. Occurrences of this heavy element are often associated with non-marine strata of Mesozoic age, strata that can also hold significant numbers of vertebrate fossil remains. Extensive field work in the western U.S. on both private (large ranches) and public (BLM) land acquainted the author with the local abundance of vertebrate fossils in both Mesozoic and Cenozoic strata, strata which are also uranium bearing and which were the target in the commercial exploration for that radioactive element.

The success of vertebrate paleontologists campaign against the collection of vertebrate fossils from public land necessitates the use of caution about such activity, however, on the other hand, numerous vertebrate fossils have come from from private land and have thus entered the collector's cabinet. Such fossils thus have no illicit connotation associated with them and should not be the concern of either the SVP or of SAFE.

These shark teeth were found in vast numbers in an excavation in the Dallas/Fort Worth area. The statement that "all vertebrate fossils are rare and thus should be given legal protection" is not true; regarding rarity, many vertebrate fossils like these shark teeth, can be extremely common fossils.

Excavations into this gravel bed yielded numerous fossil turtles while doing trenching for uranium bearing minerals. A dozer operator noticed them and brought them to my attention. Fossils, sometimes good ones, can turn up in unexpected places.

Excavations such as this for a shopping center exposed outcrops of soft mudstones which previously did not exist. Such outcrops have a potential for producing significant "new" fossils and well as giving new information on local geology.

A considerable variety of fossils and fossil related items such as books and apparel is offered at MAPS EXPO, an exclusively fossil related show given in the Spring at Macomb, Illinois. Similar shows, but with minerals and lapidary items as well as fossils, are given in many communities throughout the U.S.

Mesozoic vertebrate fossils from other parts of the world have, over the years, entered the fossil market and it might be mentioned that they usually offer a win-win situation to all involved. Examples to date are the numerous Late Cretaceous (Maastrichtian) shark, crocodile, turtle, and mosasaur fossils collected from phosphate mines in Morocco. Such fossils, by being locally collected, are being salvaged from the rock crusher and sold by miners and locals to Moroccan fossil dealers where they are then distributed worldwide. By this activity, the local economy gains a by-product from phosphate mining that produces enhanced value that goes into the local area. The abundance of fine vertebrate fossils thus salvaged enables collectors, both institutional and private, to obtain some nice material from these late Mesozoic beds.

Vertebrate fossils of Mesozoic age that have come onto the fossil market from China during the past twenty years seem to offer a similar situation, however some of these fossils may also be more controversial. It should be noted that the Chinese government's position on vertebrate fossils at the time of this writing is to allow those fossils which are of frequent occurrence, such as many of those of Liaoning Province, to go into the fossil market. Regarding the matter of the legality of the numerous Chinese dinosaur eggs, which have entered the fossil market, the author has found to be a more "muddy issue." However, from the sheer number of such eggs seen at the Tucson show during the past few years as well as on the Internet, it appears that such material occurs in sufficient abundance to reasonably allow eggs to go onto the fossil market. How many duplicates of even a unique fossil like a dinosaur egg does a museum really need?

After a reasonably quantity of such a fossil has been acquired, it appears that greater benefit to all concerned is to disperse the fossils, so that both institutions and private individuals can become familiar with what is otherwise a rare fossil. Also, with this arrangement, a sort of cross-fertilization occurs where all interested parties become familiar with and acquire material from other parts of the world. There is no better method of becoming familiar with a broad spectrum of fossils than to work with actual specimens, either in the field or in the cabinet. The attitude of "we have unique fossils and we want **all of them** to stay right here" runs contrary to the dissemination of knowledge, which a broad spectrum of fossils is capable of promoting.

Collecting Mesozoic Fossils-II

Various land managing agencies have justified fossil and rock collecting prohibitions on the basis that fossils and other geologic phenomena are a non-renewable resource, which in theory is true, but only in an abstract way. In actuality, erosion and/or other processes, which can include activities of man like quarrying and construction, are continually exposing more fossils. In soft rocks like shale, clay, and marl, this erosion may take place over short periods of time so that the weathering process destroys any fossils found in these rocks, if not collected. Such weathering, however, also keeps exposing more fossils; therefore fossils are a renewable resource on a time scale relative to humans, but yes they are a non-renewable resource with respect to immense spans of geologic time. Even hard rocks in streams or in talus (loose rock at the base of a bluff) is continually being moved and exposed, although at a rate usually not obvious on a time scale experienced by a person.

The Worldwide, Commercial Fossil Industry

During the past twenty years a variety of attractive and interesting fossils have come onto the fossil market in large quantities. These come from many countries around the world and can appeal to a broad range of persons. Many of these come through the large mineral-fossil shows in Tucson and Munich and many are from Mesozoic strata. Particularly well represented by such commercial fossils are ones from Morocco, Madagascar, China, and Russia. Commercial fossil collecting, on a large scale, is particularly strong in Morocco where an astonishing number of fossils representing a wide range of geologic ages are offered. These Moroccan fossils range from Precambrian to Cenozoic, with the majority being from the Paleozoic (trilobites, ammonoids, and nautaloids) and the Mesozoic (ammonites, mosasaurs, plesiosaurs, crocodiles, and vertebrate teeth). This activity has been criticized by some paleontologists, but especially with the Paleozoic fossils; such large scale collecting has produced a fair amount of material new to science. Specifically some of these fossils have been peculiar fish, trilobites with eye shades, machaeridians and some Burgess shale or Edicarian-like fossils known as *Eldonia*. Of the Mesozoic fossils from Morocco, ther presence of mosasaur and plesiosaur material, sea turtles, and a vast number of ammonites including heteromorphs, should be mentioned. With these, probably the greatest gain for paleontology has been direct, hands on contact with what otherwise are fossils which would not be widely seen but which can offer, to the interested person, an opportunity of ownership at little cost. Availability of these fossils for educational purposes, where they increase interest in these fossil groups, is an activity that the author considers as a benefit to paleontological education.

Value Range (Mesozoic Fossils)

The matter of placing monetary value on fossils can (and is) a sticky one and with Mesozoic fossils, maybe particularly so. Mesozoic fossils are some of those that the public is particularly attracted to and which may lend themselves particularly well to museum display. Interest in Mesozoic fossils also have soared since the *Jurassic Park* movies, books, and accompanying promotions so that persons who before *Jurassic Park* "events" wouldn't have looked twice at a fossil now want a dinosaur bone or an ammonite fossil as trendy home décor. Such a "bubble" of interest is a ephemeral thing that can give rise to inflated prices; however, the fact remains that Mesozoic fossils can have "top billing" when it comes to the public and thus can have considerable market value, often overshadowing that of other fossils. Some Mesozoic fossils, such as those from Solnhofen, Germany, have been on the fossil market for over two centuries and their values have been well established, although they can fluctuate with economic conditions. Such fluctuation with economic conditions (or rather non-fluctuation of value, it's the value of currency which often fluctuates) makes them (particularly the fine ones), fairly good investment vehicles (or at least as good as you can get with investments in fossils). Other European fossils, such as ammonites, seem less likely to hold their value over time. This is, in part, because similar fossils occur in other parts of the world and for someone wanting nice ammonites, these can be available at a fraction of the cost of what similar European specimens might sell for.

Offerings of a considerable variety of fossil and mineral specimens takes place at shows like this. Such shows offer a significant amount of "cross-pollination" in the world of both fossil and mineral collecting.

Nicely organized group of fossil echinoderms are offered here at Maps EXPO. This is one of the largest (exclusively fossils) shows in the world. A vast variety of fossils from all over the world are offered for trade or cash in this yearly exhibit.

A case in point are the ammonites currently coming from the island of Madagascar, which came onto the fossil market around 1993. Most of these are similar to European specimens but were sold at the Tucson Show at prices a fraction of that asked for European specimens. While a comparable European ammonite may sell for $50, one from Madagascar went for $3. True, European fossils can have a cachet of desirability for historic reasons, but the price disparity is such that, for a person who wants a nice ammonite at a low price, the Madagascar specimens are the way to go.

High End "Snooty Fossils"

Mesozoic fossils can also bring up the matter of large, spectacular, and well prepared fossils that sometimes can command six figure monetary values. Such fossils appeal more as decorative items than as vehicles that might be used to incite curiosity relating to science (although it is to be hoped that the latter might also be involved). Such high-end fossils can be found in the private homes of the "well to do," in casinos, and in high-end hotels and well as other places that want a cachet of "class and uniqueness" which such "snooty" fossils can convey to the public. Some paleontologists are appalled at this use of fossils; however it is predominantly the monetary incentive that initiates the quarrying of suitable fossil bearing strata in order to obtain such fossils. A by-product of such activity is the uncovering of many more modest specimens, some of which can be new and significant to science; many others uncovered by such "fossil mining" will also find their way into the cabinets of both institutional and private collectors. It is hoped that when such "high-end" specimens are no longer desired, they will be donated to some institution such as a natural history museum as a tax write off and not end up in a landfill, as did some marble containing unique fossils (archeocyathids) of which the author is aware.

The Value of
Fossils and Value Guide

The uninformed, seeing monetary values placed upon fossils, believe that, if you put in the right kind of effort, you can find them yourself, and are attracted to fossil hunting as a perceived source of easy money. Nothing could be further from the truth. Yes, one can often find fossils, but it's not like finding money. What has to be considered is the completeness, rarity, and appearance of the fossil. Fossils can be common, some whole layers of rock are entirely composed of them but these generally have no commercial value. What also has to be considered is the effort and time invested in their preparation, the removal of excess rock, and the exposure of the fossil. With many fossils, including many Mesozoic fossils such as ammonites, such "value added" labor as polishing has to be considered in a fossil's value and price. Most fossils don't look like much in the field; they have to be brought out (prepared) from the rock by painstaking manual methods or by technology using such devices as pneumatic chisels, air abrasive machines, acids or other chemicals. Fossil preparation strategies are varied and the list is extensive, as fossil preparation is a field in itself. It should be pointed out that most of the specimens in the two Mesozoic books by the author, of which this is the first, fall into the E through G ranges and that this is **after** preparation, which is usually time consuming.

The value range for fossils illustrated in this book:

A: $1,000-2,000
B: $500-$1,000
C: $250-$500
D: $100-$250
E: $50-$100
F: $25-$50
G: $10-$25
H: $1-$10

Bibliography

Dixon, Dougal. 2006. *The Complete Book of Dinosaurs*. Hermes House, Anness Publishing, Ltd., London.

Gould, Stephen J. 1985. "Sex, Drugs, Disasters, and Extinction of Dinosaurs," in *The Flamingo's Smile: Reflections in Natural History*. W. W. Norton and Co., New York-London.

_____.1995. *Dinosaur in a Haystack. Reflections in Natural History*. Harmony Books, New York-London.

Lyell, Charles. 1833. *Principles of Geology*. John Murray, London.

Ward, Peter. 1983. "The Extinction of the Ammonites." *Scientific American*. No. 249.

Eon	Era	Period	Age my.
PHANEROZOIC	Cenozoic	Tertiary	0 / 67
	Mesozoic	Cretaceous	
		Jurassic	
		Triassic	235
	Paleozoic	Permian	280
		Pennsylvanian	
		Mississippian	300
		Devonian	350
		Silurian	320
		Ordovician	440
			500
		Cambrian	
			542
		Precambrian ↓	

The Phanerozoic geologic time scale. The Cretaceous-Tertiary boundary (K/T boundary) is at 67 million years ago, a date that has been well established through radiometric age dating.

Era	Period	Age my.
Cenozoic	Tertiary	0 / 67
Mesozoic	Cretaceous	
	Jurassic	
	Triassic	235
	Permian	280
	Pennsylvanian	
	Mississippian	

The Mesozoic Era with the Triassic (red), Jurassic (yellow), and Cretaceous (green) indicated by conventional colors of these periods as used on geologic maps.

Chapter Two
The Triassic Period

During this part of geologic time, life takes on a new direction, a direction which is quite different from that of the Paleozoic Era. The extinction of most Paleozoic life "cleared the stage" for the next "act," an "act" which would find many (mostly) "new" players.

Land Plants

A dominance of oxidized sediments (red beds) in the Triassic suggests widespread desert conditions, an environment not very conducive to plant life, particularly as desert plants like cacti and succulents are geologically recent additions to the Earths flora (they are specialized angiosperms). Those plants that existed in the dry uplands of the Triassic (and earlier) remain a mystery. Swamp-like environments did exist in the Triassic and it is here that fossil plants (and even coal formation) have left a paleobotanical record. Most of the Triassic plants represented here are from such a "coal swamp" environment.

With Triassic plants the tree ferns were supplanted by conifers, cycads, and the bennetailes; however, vestiges of late Paleozoic plants like *Neocalamites* and *Pecopteris* do occur.

Pecopteris sp. This is an example of a "hold-over" from the Paleozoic Era. *Pecopteris* is a very common fern in late Paleozoic rocks (Permian and Pennsylvanian periods). A few Paleozoic genera like *Pecopteris* carried over into the Triassic but for the most part (some 95+ percent) of Paleozoic life forms went extinct at the end of the Permian in what is considered to be the earth's greatest extinction event; it is known as the terminal Paleozoic extinction event. These fronds are preserved as a carbon (graphite) film. This and some of the following photos from the Santa Clara Formation have been taken in low angle light and contrast has been enhanced. Santa Clara Formation, Tarahumara, Sonora, Mexico.

Same specimen as above but without photographic enhancement.

A closer look at a Triassic pecopterid with photographic enhancement. Santa Clara Formation. Mexico.

Cladophlebis australis. Cladophlebis is a widespread fern of the Triassic and Jurassic. So abundant locally was this fern that Triassic coal beds in eastern Australia and Tasmania have formed from vast accumulations of it in coal swamps. Triassic fossils tend to show up less frequently on the fossil market then do those of other ages and fronds of *Cladophlebis* should be more abundant then they are as they occur frequently in Australian coalmines. Merrwood Coal Mine, Royal George, Tasmania. Triassic coal is uncommon and fossil plants found during mining should be dispersed, as they are scientifically and educationally valuable.

Ctenophyllum (Cladophlebis) branianium. This fern is associated with the foliage of cycads and cycad-like plants (bennitailes). Like the tree ferns of the late Paleozoic, Triassic ferns often were large plants. Low angle light reflecting off of the graphite film of the frond as well as increased contrast has enhanced the appearance of this specimen. Late Triassic (Carnian), Santa Clara Formation, Tarahumara, Sonora, Mexico. (Value range F).

Ctenophyllum (Cladophlebis) branianium. Iron stained pinnate leaves of this Late Triassic fern are preserved in black, slaty shale. Specimen has been photographically enhanced as black on black fossil leaves are often hard to see. Late Triassic, Santa Clara Formation, Tarahumara, Sonora, Mexico.

Ctenophyllum sp. A larger leaf than shown in previous photos of this Triassic genus. Similar ferns have been found in shale beds of the Chinle Formation associated with the Petrified Forest of northern Arizona. The Santa Clara Formation of Sonora probably correlates with the Chinle Formation of Arizona. Specimen has been photographed in low angle light and has been photographically enhanced. (Value range F).

Same specimen as above but with no photographic enhancement.

Another specimen of this fern from the Santa Clara Formation. Upper Triassic, Sonora, Mexico.

Same specimen as shown above but with no low angle light or photographic enhancement.

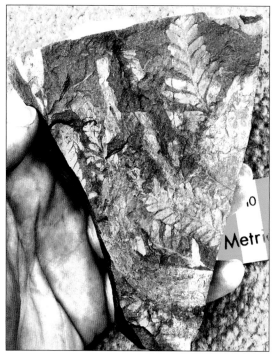

A slab with *Ctenophyllum* which has been photographically enhanced. Santa Clara Formation, Sonora, Mexico.

Calamites (Neocalamites) arenaceous. The late Paleozoic arthrophyte *Calamites* is represented in the Triassic by a holdover quite similar to the Paleozoic genus *Calamites*. These are distinguished from their Paleozoic relatives by the Triassic forms being given the genus *Neocalamites*. *Neocalamites* is generally smaller than was the Paleozoic *Calamites*. Lower Keuper Formation, Upper Triassic, (Werksandsteinbruch) near Wurzburg, Germany. (Value range G).

Pterophyllum spatulatum. These cycads come from Triassic strata deposited in a rift zone (Durham Basin) in eastern North Carolina. Triassic rocks of the southern Triassic outcrop area of eastern North America contain fossil plants while similar age strata to the north are noted for their tracks and trackways of dinosaurs. Dunker Formation, Durham Basin, near Durham, North Carolina. (Value range F).

Pterophyllum spatulatum. Same specimen as above but with different lighting and more contrast. Dunker Formation, Durham Basin, Durham, North Carolina.

Pterophyllum sp. A short section of a cycad frond. Many of the cycad fossils of the Triassic and Jurassic are believed by paleobotanists not to be true cycads but rather to have belonged to an extinct division of plants known as the Bennetailes. Unless reproductive structures are preserved with the leaf compressions, distinguishing between the two groups is seemingly impossible. Triassic rift zone sediments, Durham Basin near Durham, North Carolina. (Value range F).

Podozamites sp. Iron stained leaves of this "cycad" are different from the usual "palm-like" leaves of cycads. Pechgraben, Kulmbach, Germany. (Value range F).

A large "cycad" leaf characteristic of Triassic strata worldwide. Like other "cycads" of the Mesozoic, it has been questioned whether it is a true cycad as many Mesozoic "cycad-like" leaves can belong to the Bennitailes, an extinct group of plants that had a reproductive structure (cone) different from the cycads. Santa Clara Formation, late Triassic, Tarrahumia, Sonora, Mexico. (Value range G).

Close-up of leaf arrangement of *Podozamites*. The Triassic of Germany has yielded a variety of interesting and often puzzling fossils that include plants. Pechgraben, Kulmbach, Germany.

Zamites sp. This is probably a true Mesozoic cycad. Santa Clara Formation, Late Triassic, Tarahumamia, Sonora, Mexico.

Macrotaeniopteris sp. This Triassic fossil plant, unlike the above specimen, is questioned as to its being a true cycad. Many Triassic "cycads" belong to the Bennitailes, an extinct cycad-like plant division. *Macrotaeniopteris* is a distinctive and typical Triassic plant. Santa Clara Formation, Late Triassic, Tarahumara, Sonora, Mexico.

Macrotaeniopteris sp. The same specimen as in the previous photo but with contrast somewhat enhanced. Separations in these distinctive Triassic leaves can be seen more distinctly here. Fossils such as these black on black leaves are often best seen by reflected light, which can then be photographically enhanced. (Value range F).

Araucarioxylon arizonicum. A petrified "round" of what is thought to be an early sequoia-like conifer. This is the designation of most of the Triassic fossil wood found in northern Arizona including much of that found in the Petrified Forest. Such petrified wood is highly silicified with various forms of iron oxide providing a red or yellow pigment. Conifers are well represented in the Triassic by fossil logs like this. Chinle Formation, Upper Triassic, Holbrook, Arizona.

"Araucarioxylon." A sliced section of highly silicified wood from the Chinle Formation near Holbrook, Arizona.

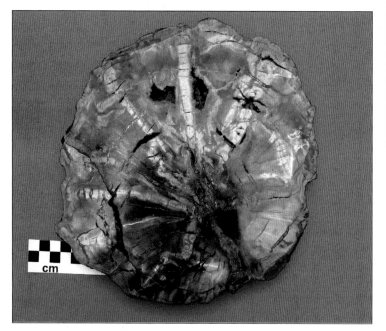

Woodworthia sp. *Woodworthia* exhibits a star-like pattern in a log cross-section. The Chinle Formation carries petrified wood in many of its outcrop areas in Arizona, Utah, New Mexico, and Colorado. *Woodwoorthia* is generally found in the Chinle Formation of southern Utah while much of the wood of northern Arizona belongs to *Araucarioxylon. Woodworthia* is believed to represent a type of sequoia-like conifer, but with a buttress-like base like that of the Bald Cypress, Chinle Formation. Upper Triassic, southern Utah. (Value range F).

"Araucarioxylon." Slabs of petrified wood and the highly silicified log section of the previous picture come from the Upper Triassic Chinle Formation of northern Arizona. The largest concentration of this wood is in the Petrified Forest and Painted Desert of northern Arizona. This area was given National Monument status in the early twentieth century, partially because an abrasives company had filed mineral claims in 1906 on the area with the intent to crush and pulverize the logs for use as abrasives in sandpaper. Such a "Philistine" mindset still can prevail in some quarters of the mineral industry where unique and desirable specimens are sent to the rock crusher as it is too much trouble to bother saving them. Chinle Formation, Upper Triassic, Holbrook, Arizona.

Large petrified logs of what is probably Sequoia "Araucarioxylon" from the region of the Petrified Forest in northern Arizona. *Photo courtesy of Warren Wagner.*

Petrified (silicified) logs in the Petrified Forest of Arizona. Most of these silicified logs, which have weathered from the Chinle Formation, belong to sequoia. *Photo courtesy of Warren Wagner.*

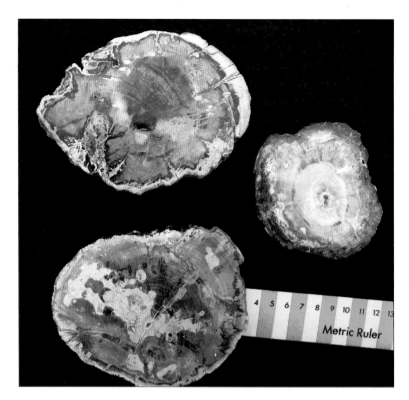

"Araucarioxylon." Triassic strata of Madagascar locally yield concentrations of petrified wood similar to that of the Chinle Formation of Arizona and Utah. Shown here are sliced and polished sections of conifers, probably sequoia. Like Triassic petrified wood of the Chinle Formation, this also is quite colorful. (Value range F, single polished slice).

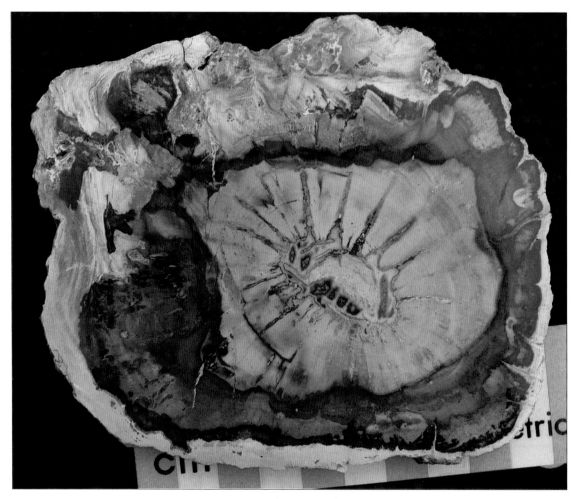

Woodworthia sp. *Woodworthia* is a Triassic conifer which has a characteristic star-like pattern in the center of a "round." Similar petrified logs occur in the Triassic Chinle Formation of southern Utah. *Woodworthia* is believed to be the wood of a relatively large conifer. Triassic of Madagascar. (Value range F).

In the animal kingdom, many lower invertebrate animals such as corals are of a "new" type; these new type of corals are known as hexacorals and appear after the Paleozoic extinction event. Brachiopods, that invertebrate mainstay of shallow Paleozoic seaways, never again reached anything close to the dominance that they had during the Paleozoic Era. Sponges, the lowest of invertebrate "animals" (if they are considered animals at all), stayed pretty much the same as those found in the Paleozoic.

(Sponge). A silicified sponge from Triassic rocks of the Dolomite Alps, which form the border between Italy and Austria. Sponges and other fossils, like corals, can be silicified like this specimen where they then weather from the limestone (or dolomite) in relief. Other marine invertebrates, like corals and brachiopods, can occur with these sponges but Triassic versions of these fossils are considerably rarer than are those of the Paleozoic Era. Muschelkalk, Middle Triassic, southern Austria. (Value range F).

Encrinus liliiformis. A crinoid which can compose "crinoid gardens" in the Middle Triassic Muschelkalk Formation of southern Germany. A large stem is on the right side of the slab with a holdfast at its base. These were some of the first fossil crinoids to be investigated by science, an event that took place in the seventeenth century. They were described as puzzling phenomena and referred to as "wheel stones." Stem fragments were specifically known by the Latin term of *trochites* and sections of the crinoid column as *radersteine*. Crinoids, as a distinct group of organisms, were not recognized until the late eighteenth century and fossils of them were not recognized as such until the early nineteenth century. Muschelkalk Limestone, Crailsheim, Germany.

Echinoderms

Echinoderms, as in the Paleozoic Era, are well represented in the marine fossil record, however a number of Paleozoic classes of this group are now extinct, like the blastoids and cystoids. New forms become abundant, like the heart urchins, which are well represented in the seas of today. Crinoids that are related more closely to those living today replace the various archaic crinoid orders of the Paleozoic.

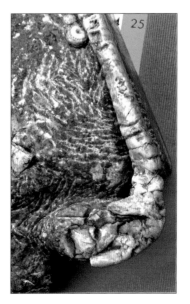

Close-up of the bottom of the large *Encrinites* stem of the previous photo. The bottom portion, where the crinoid was attached to the sea floor or to some other object, is referred to as a holdfast. The holdfast of this crinoid resembles a golf club (driver). Muschelkalk Formation.

Crinoidal Muschelkalk Limestone. Crinoids, when they die and fragment (disarticulate), can accumulate (if present in large numbers) into what is known as crinoidal limestone. Few or no recognizable crinoids as to genus or species will usually be observed in such organic limestone.

Encrinus lilliformis Two heads (calyx's) of this distinctive German Triassic crinoid. Muschelkalk Limestone, Crailsheim, Germany.

Two of the slabs bearing the crinoid *Encrinus lilliformis*.

Trarematocrinus sp. This group of fine crinoids came from a crinoid "reef" of about the same age as the Muschelkalk Limestone of Germany. All of these crinoids were living in the waters of the Tethy's Ocean, which was the single large ocean that existed at the same time as the super continent of Pangea. From Middle Triassic Limestone, Xinyi, Guizhou Province, China.

Trarematocrinus sp. A large slab of Middle Triassic Limestone bearing superb specimens of crinoids from Xinyi, Guizhou Province, China. (Value range B).

A closer view of one of these large, spectacular slabs from China, which showed up at the Tucson, Arizona, show in 2001.

Trarematocrinus sp. A large, splayed calyx of this spectacular crinoid from the Middle Triassic, Xinyi, Guizhou Province, China. (Value range C).

The calyx of one of the large Middle Triassic crinoids from Xinyi, Guizhou Province, China.

Lenticidaris utahensis Kier. A cidarid echinoid from a critical part of earth history, the Lower Triassic! The Permian extinction event decimated most invertebrate groups to the point that early (lower) Triassic rocks are generally quite sparse in fossils. The cidarid echinoids had a Paleozoic presence and escaped extinction to diversify again in the Mesozoic Era. These sea urchins came from the Lower Triassic Moenkopi Formation of the American Southwest. The Moenkopi is generally sparse in fossils, even though the formation is a Triassic extension of the underlying fossiliferous Kiabab Limestone of Late Permian age. Lower part of the Moenkopi Formation, St. George, Utah.

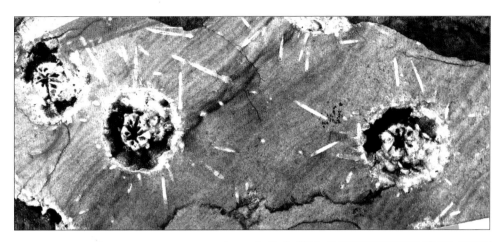

Two cidarid echinoids which have been exposed by slicing the echinoid bearing limestone layer with a thin diamond saw blade. Moenkopi Formation, Lower Triassic, St. George, Utah. (Value range F).

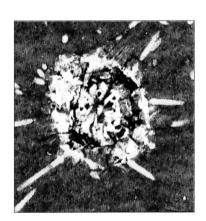

Close-up of a group of specimens of the echinoid *Lenticidaris utahensis* showing the centrally located Aristotle's lantern, a unique anatomical element characteristic of echinoids. Moenkopi, Lower Triassic, near St. George, Utah.

Aspidura scutellata Blumenbach. These minute starfish (asteroids) are associated with algal "blobs" of the Middle Triassic, Muschelkalk of southern Germany.

Aspidura scutellata. Close up of these very small asteroids, they are between the red dashes. Muschelkalk, Schwabish Hall, southern Germany. (Value range F).

Mollusks

Mollusks are represented by pelecypods and gastropods, also by cephalopods in the form of the ammonoids (ceratites and ammonites). The ammonoids undergo great taxonomic radiation and then some go extinct: ceratites at the end of the Triassic, ammonites at the end of the Mesozoic Era. A sequence of cephalopods classified on the basis of chamber-wall sutures has as its most "primitive members" the nautaloids with their plain, watch-glass-like suture walls. All other shelled cephalopods with sutured chamber walls (which include the ammonites) are known as ammonoids. Next in the series after nautaloids are goniatites, ammonoids with "jags" in their chamber walls. These are well represented by the Devonian ammonoids of Morocco. Next in suture complexity come the ceratites, an ammonoid with greater suture complexity than a goniatite. Ceratites are predominantly Triassic fossils and go extinct at the end of that period with the Triassic extinction event. Ammonites, those shelled cephalopods with the greatest structural complexity, are best represented in the Jurassic and Cretaceous where they go extinct at the end of the Cretaceous.

Tropites sp. An evolute ceratite. Ceratites are ammonoid cephalopods with a suture pattern more complex than that of a goniatite but less complex than that of an ammonite. Evolute ammonoids show the previous coils of the shell. This ceratite is from red limestone of Timor, Indonesia, where fossiliferous Permian through Triassic strata yield a variety of sometimes unique fossils. (Value range F).

Propanoceras sp. An involute ceratitic ammonoid from Triassic strata of Indonesia. (Value range F).

Involute Triassic ammonoid (right), evolute Triassic ammonite (left). Permian strata of the island of Timor grades into Triassic strata so that what are known as Permo-Triassic strata span the Permian extinction event. The upper layers of this Permo-Triassic sequence carries involute ammonoids like this specimen, an ammonoid characteristic of the Triassic. Island of Timor, Indonesia. (Value range F).

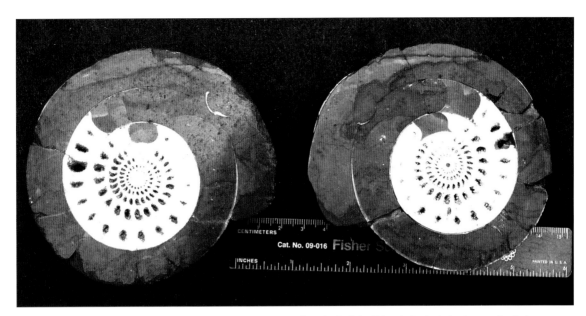

Cladiscites sp. An involute ammonite (shell coils inside itself) typical of the Triassic Period, the inner whorls have been filled in by white calcite. Distinctive red limestone of Austria, Northern Italy, and Greece yield a variety of unusual ammonites like this specimen. A region around Hallstadt, Austria, has been particularly well known for them. Upper Triassic, Hallstadt, Austria. (Value range E, both sawed halves).

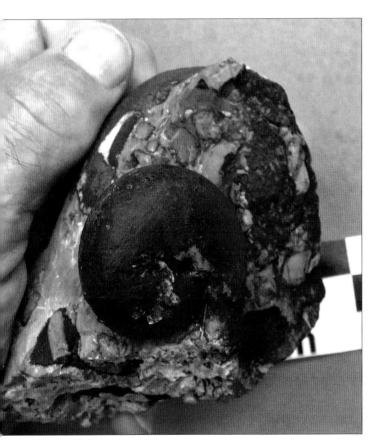

Arcestes (*Johanites*) sp. This small ammonite comes from red Triassic limestone deposited in the Tethy's seaway of southeastern Europe. *Arcestes* has a distinctive "hooded" aperture that, however, is not preserved on this specimen. Argolis, Greece. (Value range F).

Arcestes (*Johanites*) sp. This involute ammonite is characteristic of the mid- and upper Triassic. The specimen comes from a distinctive red limestone, which outcrops in Austria, Northern Italy, Yugoslavia, and Greece. Middle Triassic, Argolis, Greece. (Value range F).

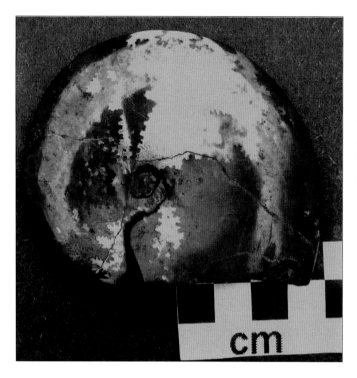

Parapinacoceras aspidoides (Diener).
A very thin (and rare) ammonite genus
from the Triassic red limestone of
southeastern Europe. Middle Triassic,
(Ladinian) Dolomite Mountains,
northern Italy. (Value range D).

Mojsvarites agenor (Munster). A
Ceratite. Ceratites represent an order
of ammonoid in which the sutures are
less complex than are those of most
other Mesozoic ammonoids. There
is a sequence of progressively more
complex sutures represented by the
nautiloid-goniatite-ceratite-ammonite
sequence. This specimen came from red
Triassic limestone of the Dolomite Alps of
Northern Italy. The Dolomites form part of
the boundary between Italy and Austria.
Middle Triassic. (Value range E).

Proarcestes bramanteri (Mojsisovics). A small,
involute ammonoid from Ladinian (Middle Triassic)
red limestone of the Dolomite Alps, Udihe, northern
Italy. These ammonites, found in red limestone from
northern Italy and Austria, usually have to be polished
to bring out their diagnostic sutures. They are also
relatively expensive when available. (Value range E).

Meekoceras gracilitatus. Involute Lower Triasssic ammonoids from fossiliferous beds of Nevada. Ranges in the Great Basin of Nevada are famous for their excellent Triassic ammonoids (ceratites and ammonites). *Meekoceras* is an example of an invertebrate that proliferated after the Permian extinction event; most other invertebrates did not become abundant in large populations (capable of making fossil beds) until the mid- or late Triassic. Lower Triassic, Scythian, Elko County, Nevada.

Paraceratites cricki (*Ceratites spinifer*). Ceratites are characteristic ammonoids of the Triassic. Ammonoids of this type, originally known as *Ceratites spinifer*, have been known and collected from the Humboldt Range of Nevada since the late nineteenth century. These ammonoids are examples of a Ceratite, an ammonoid order that goes extinct at the end of the Triassic Period. Prida Formation, Fossil Hill, American Canyon, Humboldt Range, Nevada. (Value range F).

Paraceratites cricki. A number of species and genera have been split off of the ammonoid *Ceratites spinifer* from the Humboldt Range of Nevada. Prolific fossil localities and formations like the Prida Limestone tend to promote paleontological "splitters." The Prida ammonoids occur in a concentrated manner, most of which have been described in earlier literature as *Ceratites spinifer*. The author questions the validity of many of the genera and species that have been split off from *C. spinifer*. Taxonomy is a subjective matter but given enough fossils and time, splitters seem to prevail. In most populations of animals, living together of closely related species is not the norm, however there are exceptions; splitters seem to assume that such exceptions are the norm. Prida Limestone, Middle Triassic, Humboldt Range, Nevada.

Ceratites spinifer Smith. A group of various sized specimens from Fossil Hill, west Humboldt Range, Nevada. These specimens came from an old collection and were originally sold by Ward's Scientific Co. of Buffalo, New York, sometime in the 1910s or '20s. Commercial collecting, like that done previously by Wards, is now illegal on public land in the U.S. such as that of the Humboldt Range. Commercial collecting gets specimens into the hands of those interested in them but can also infringe on an individual's ability to collect and do geology as such activity may remove most of the accessible fossils. Prohibition of commercial collecting, while allowing individuals access to casual collecting, seems to be a reasonable compromise to what can be a divisive and "thorny" issue. *Damella* zone, Middle Triassic, Fossil Hill, American Canyon, West Humboldt Range, Nevada. (Value range G, individual specimen).

A small ammonite from the Prida Limestone, Humboldt Range, Nevada.

Arthropods

With the arthropods, lobsters make their first appearance in the Triassic Period; trilobites, those other major arthropod players of the Paleozoic Era, became extinct at the end of the Paleozoic Era. New insect families appear in the Mesozoic, however Triassic rocks are usually reluctant to give up their insect fossils.

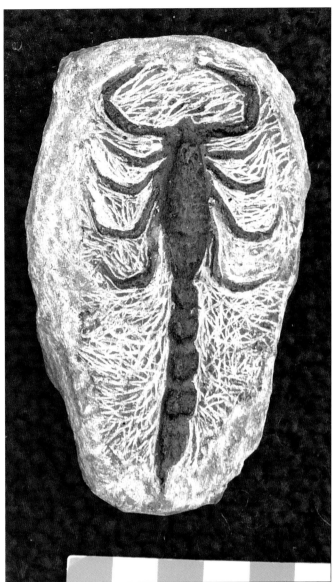

Pexphix severi. An early lobster. Lobsters and (hard shelled) crabs enter the "picture" in the Mesozoic Era. This is an early lobster from the Middle Triassic Muschelkalk Limestone of southern Germany. Grailshiem, Brunswick, Germany.

"Scorpion." This "fossil scorpion" is a fake! These are made by Moroccan fossil perpetrators who want to be creative. This fake fossil is almost identical to carved, fake fossils described and illustrated in a famous book published in 1726 by Johannes Beringer of Wurtzberg, Germany. (See the *Lying Stones of Professor Beringer* or "The Lying Stones of Marrakech.") This scenario came about through Professor Beringer's taking students to a fossil bearing, Triassic shale outcrop (Keuper Formation) near Wurtzberg to collect fossils. One of these students carved a "fossil" in the shale that the professor found with delight. Soon many more fabulous fossils like this scorpion turned up in the talus, which were also found by Beringer. Professor Beringer became so excited with these finds that he wrote and published an extensive work on them in 1726, fossils being a topic that was still puzzling and problematic at the time. His book, with their numerous woodcuts, showed not only scorpions but "fossil" toads, insects, and snakes. Only after he started finding specimens with his name on them did he realize that he had been duped. To ease his embarrassment, he tried to buy back and destroy all of his books on the fake fossils; fake fossils that he now realized had been carved as a prank by his students or by personal enemies. (The story is actually a bit more complicated but that outlined here is what is usually offered.) Both his books and the fake fossils on which the book was based became collectable and insured the professor's fame. More accessible today are similar modern fakes from Morocco. (Value range G).

Vertebrates

It's with the vertebrates that things really get going, in the Triassic Period. Fish in the Triassic are still primarily represented by the ray-finned fish, the coelacanths and the dipnoans (lung fish). It's with land vertebrates that the real innovations occur. Reptiles of the Triassic are probably more bizarre than they were at anytime during their geologic history, with such types as mammal-like reptiles, flying reptiles (pterodactyls), ichyosaurs, thecodonts, and—in the late Triassic—dinosaurs being the main "players." The Triassic ends with a major extinction event in which members of some invertebrate groups, like the ammonites, become considerably reduced and others, like the ceratites, go extinct. Some reptiles, also characteristic of the Triassic, such as placodonts and the thecodonts, go extinct at the end of the period as well as do many amphibians like the stegocephalians.

Paleoniscus sp. A "ray-finned" fish from the Middle Triassic Muschelkalk Limestone of southern Germany.

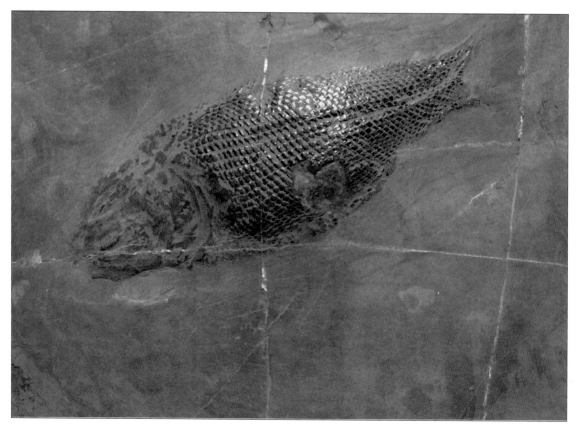

Sinuengnathus shynglensis. A paleoniscid fish from Triassic slate. The somewhat metamorphosed strata that yielded this fish also produced numerous small reptiles (*Keichosaurus*). Huxia Formation, Guarglin, Guizhou Province, southwest China.

Parieidus madagascarious. Part and counterpart of a full-bodied, early ray-finned fish preserved as an impression in an ironstone nodule. Such nodules formed around the fish in a manner described by Hugh Miller of Devonian "Old Red Sandstone" fame as a "stony sarcophagus." In the Madagascar nodules, the fossil fish is not entirely flattened but has left a 3-dimensional impression where the once-present fish has left a cavity. Such nodules weather from shale beds and split open where the fossil forms a plane in the nodule. These Triassic fish bearing nodules (or concretions) are some of the nicest of such fossils to come recently (2004) onto the fossil market. Lower Triassic, Ambilobe, Madagascar. (Value range E for particularly complete specimens like this).

Two fish bearing nodules with part and counterpart. Lower Triassic, Ambilobe, Madagascar. (Value range G, single nodule-part and counterpart).

Ray-finned fish. A superbly preserved ancestor of modern fish (Teleosts) from the Middle Triassic Muschelkalk of Hallstadt, Austria. (Value range D).

A large, jumbled (disarticulated) fish in an ironstone concretion. The fish responsible for this nodule was partially decomposed and its scales and bones were scattered prior to their burial in fine mud and the formation of this concretion. The fragmented fish-remains became the nucleus around which the nodule (or concretion) formed. Lower Triassic, Ambilobe, Madagascar. Such jumbled fish are interesting but are less desirable than are complete specimens. (Value range G).

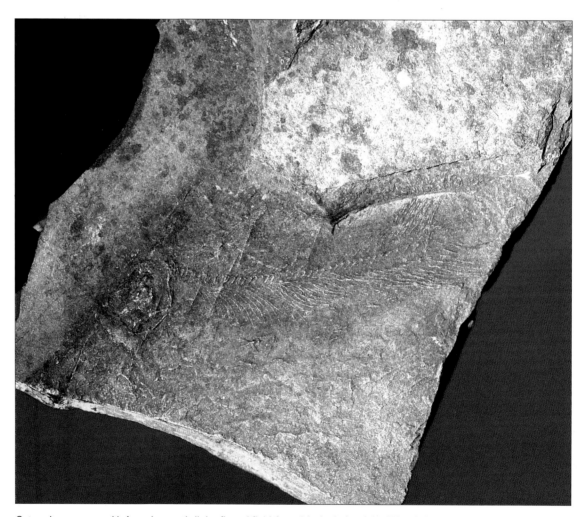

Osteopleurus newarki. A coelencanth (lobe finned fish) from black shale of the Triassic Newark Basin. A number of these fossil fish, preserved in dark black shale (which are hard to photograph), come (or came) from various excavations in the southern New York metropolitan area of New Jersey. One of these localities, ironically, was the excavation for the Biology Department building at Princeton University in Princeton, New Jersey. Specimen from the Lockatung Formation, Guttenburg, New Jersey.

Leptolepis tallragerensis. An early bony fish. Bony fish or teleosts are the predominant fish living today in both fresh and salt water. This is one of the earliest occurrences of a teleost! These early bony fish are associated with fossil plants of the *Glossopteris* flora, a distinctive flora characteristic of the Permian and Triassic of the southern hemisphere (Gondwanaland). Upper Triassic, New South Wales, Australia. (Value range F).

Stegocephalian. A replica of a head shield of a Triassic amphibian (cast). These large, bone-headed amphibians were denizens of Permian fresh water lakes and streams and survived the Permian extinction event but then went extinct at the end of the Triassic Period, specimen approx. 12 inches in length. Lower Triassic, Dockum Formation, west Texas.

Stegocephalian amphibian fragments. Complete Stegocephalian shields are rare fossils. What are usually found are shield fragments like these. The large plate is from the Chinle Formation, Valicito, Colorado; other fragments and non-descript bones are from the Dockum Formation of west Texas. (Value range E for group).

Pachypleurosaurus (Neusticosaurus) edwardei. A small lizard-like reptile found in Triassic strata of Italy and Switzerland (Lake Constance). Specimens of this saurian are preserved in brackish water deposits of a Triassic lake. These fossils are similar to *Keichiosaurus* from China. This is a plaster cast of what would be a fairly pricey fossil if original. Replica from Jones Fossil Farm, Worthington, Minnesota. (Value range F).

Keichousaurus hui. A medium sized specimen of this marine "saurian" (lizard) from late Triassic strata of Southwest China. This specimen exhibits a number of small cracks that are filled in with quartz. Quartz veins like these form an unusual combination of deep-seated geology (mineral formation) and vertebrate fossils. Quartz veins can be associated with mineralizing solutions associated with tectonic activity. Quartz veins, even small ones like this, are also generally associated with a type of geology usually thought of as having nothing to do with fossils, much yet vertebrate fossils like this saurian. A number of these interesting fossils came onto the fossil market in 2004. Upper Triassic, Huixia beds, Xingyi, Guizhou Province, Southwest China. (Value range E).

Same specimen as in the previous photo, different view and lighting.

A relatively large *Keichousaurus* specimen with numerous quartz veins These are some of the highest quality articulated fossil vertebrates to come onto the fossil market, particularly considering the relatively large numbers of specimens available. (Value range D).

Keichousaurus hui. A small, slightly damaged specimen of this Triassic saurian. Guzhou Province, Southwest China. (Value range F).

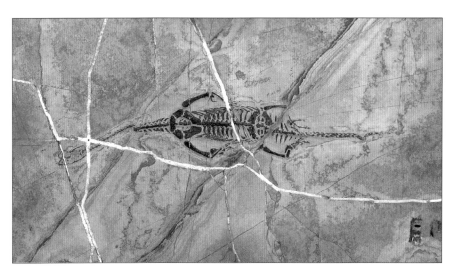

Keichousaurus hui. A larger, complete specimen of this Triassic reptile preserved in slaty shale. Note the narrow quartz veins traversing the specimen, something not usually associated with vertebrate fossils but usually associated with "deep time." Huixia beds, Xingyi, (Guanglin), Guizhou Province, Southwest China. (*Specimen courtesy of Steve Riggs Jones*). (Value range D).

A large *Keichousaurus* specimen with quartz veins intersecting the fossil bearing slab. Quartz was introduced into rock bearing these fossils when it was buried 2+ miles below the earth's surface. These fossils come from a relatively deep-water environment and sediments from deep-water environments often lack abundant fossils. Deep-water sedimentary rocks buried deep enough into the earth's crust may contain quartz veins, but no fossils. These fossils are unique by being associated with these quartz veins, but even more so by being vertebrates. Quartz veins are emplaced into the crust by hydrothermal (water-heat) activity. Large quartz veins can be the source of valuable metals such as silver, copper or gold where these metals are introduced through the action of hot fluids or gases which come from the earth's interior and which form the quartz veins. The hard rock containing these reptiles has been altered by pressure (metamorphosed) as a consequence of its deep burial. Carbonaceous material in the original shale prevented the rock from becoming highly crystalline in the process of being deeply buried and metamorphosed.

Keichousaurus hui. An example of this small saurian with a broken neck. It takes a bit of reflection to realize (and empathize) that this fossil was once a small, sentient animal that met its end 220 million years ago and is not just a geologic specimen. Small quartz veins intersect the slab; these veins were produced when the preserving sediments were buried at considerable depth into the earth's crust. Xingyi, Guizhou Province, China. (Value range E).

Keichousaurus hui. A small specimen of this nice fossil vertebrate that has been widely distributed. Note where the fossil saurian has been offset in the neck where the white quartz veins intersect the saurian's body. Xingyi, Guizhou Province, China.

Close-up of quartz vein in a different specimen.

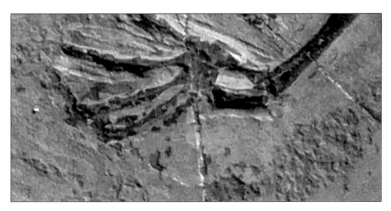

Close-up of quartz vein and Keicheosaur.

Old silver mine worked in quartz veins just west of the quartz veins shown in the previous picture. Specimens of the saurian *Keichiosaurus* and their quartz veins unite what are usually viewed as two mutually separate fields of geology, mineral genesis and vertebrate paleontology.

Hydrothermal quartz veins similar to those of the *Keichiosaurus* bearing slabs (but larger). These quartz veins, exposed in the bed of an Ozark stream were emplaced by hot fluids coming from the earth's interior. Accompanying the quartz bearing fluids was silver, lead, and gold which locally was emplaced with the quartz veins. Such veins have been heavily prospected for these precious metals and a silver mine (or prospect) occurs just west of this outcrop at the foot of Bell Mountain, Missouri Ozarks.

Phytosaur sp. These are the teeth of a phytosaur, a relatively large, crocodile-like saurian from the Triassic of the U.S. Southwest, Chinle Formation, Tucamcari, New Mexico. (Value range F).

Triassic Dinosaur Tracks and Trackways

The worlds first fossils relating to dinosaurs that caught the attention of science (although they weren't recognized as such at the time) were fossil tracks and trackways found in the early nineteenth century in reddish rock layers along the Connecticut River in Massachusetts and Connecticut. In the 1820s and '30s, a faculty member of Amherst College, Professor Edwin Hitchcock, documented many of these fossil tracks—tracks which he found in rocks exposed along the Connecticut River Valley. Believing that they were the tracks of large birds, Hitchcock amassed a large collection of them and published extensively on them. The discovery of the fact that large, lizard-like reptiles (terrible lizards or dinosaurs) existed in the geologic past was made known through Sir Richard Owen, an English anatomist in the mid-1840s; this "discovery" of dinosaurs and the concept of dinosaurs required a significant reinterpretation of Hitchcock's tracks and trackways. Dinosaur tracks and trackways are the subject matter of a branch of paleontology known as paleoichnology, the study of fossil tracks and trails. Many organisms of the geologic past have left a fossil record known only by their tracks. The goal of paleoichnology is to discover, document, and sort out the vast number of such "trace fossils" that exist; this, of course, included the tracks of dinosaurs.

What are now considered as dinosaur tracks of the Connecticut River Valley represent one of the largest concentrations of such fossil tracks in the world. Rock strata of the same age occur on the Colorado Plateau and preserve similar tracks and trackways, but in the West they are much rarer and limited in their occurrence, compared to those of eastern North America. It has recently been determined that some of the track bearing strata in the eastern states are Lower Jurassic in age rather than Triassic. Rock strata bearing these fossil tracks were deposited in what have become known as Triassic Basins (or rift zones), which formed from the opening of what would later became the Atlantic Ocean. Such basins (or "rift zones") were the cause of the subsidence, which accumulated great thicknesses of sediment in the Triassic. This sediment, however, actually continued to accumulate into the Jurassic Period. The higher (hence younger) strata found in the Triassic "Basins" are therefore early Jurassic rather than Triassic in age. Most of the tracks and trackways of the Triassic Basins are still considered as Triassic in age and that is where most of these eastern U.S. classic dinosaur tracks are placed in this book.

Natural cast of a dinosaur track from the lower part (sole) of a sandstone bed. Triassic strata of Massachusetts and Connecticut have yielded (and yield) a considerable variety of dinosaur tracks and trackways. This specimen was found in rock from an excavation near Hartford, Connecticut. Three toed tracks like this were originally thought to have been the tracks of large birds. (Value range F).

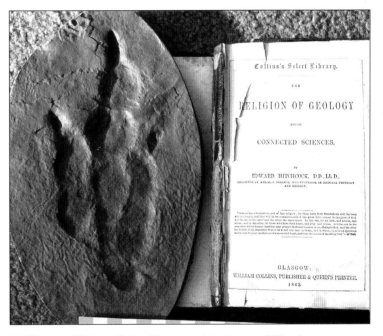

Grallator sp. (Hitchcock). Fossil tracks and trackways of Triassic (and Lower Jurassic) strata of the Connecticut River Valley were first described by pioneer geologist-paleontologist Edwin Hitchcock of Amherst College in the early nineteenth century. Hitchcock was one of the earliest workers to do extensive paleontology in North America when modern geology appeared with its concern for fossils and megatime. Hitchcock believed that *Grallator* was a crocodile-like animal; it was later interpreted as the track of a small, bipedal dinosaur. Next to the *Grallator* cast is a copy of one of Hitchcock's books, which attempted to intertwine what was being discovered in the way of fossils and rock strata with the book of Genesis of the Bible. Many early geologists, like Hitchcock, had been trained in theology and endeavored to interpret fossils and enclosing rock strata as a method to gain insight into God's Creation, such is the subject of his book.

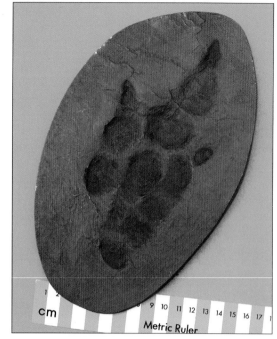

Grallator sp. (Hitchcock). Another view of a *Grallator* track from Turner's Falls, Massachusetts. This specimen (a cast) comes from the Newark Group, which is now considered as Lower Jurassic rather than Triassic in age. It is placed here, in the Triassic, as these fossils have always been considered as "Classic Triassic Fossils." The moniker "Grallator," coined by Hitchcock, sounds like it could be the brand name for a coffee grinder or a trash shredder. (Value range F).

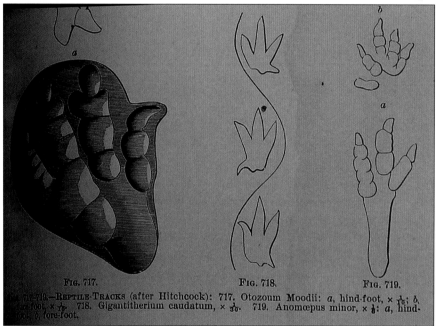

FIG. 717.   FIG. 718.   FIG. 719.

717-719.—REPTILE-TRACKS (after Hitchcock): 717. Otozoum Moodii: *a*, hind-foot, × ⅟₁₀; *b*, fore-foot × ⅟₁₀. 718. Gigantitherium caudatum, × ⅟₃₀. 719. Anomœpus minor, × ⅟₂: *a*, hind-foot; *b*, fore-foot.

Otozoum moodii Hitchcock. Hitchcock's ichnogenus *Otozoum* has been found to represent tracks (or trackways) made by a dinosaur indenting underlying layers of sediment as it walked on them rather than representing a track made directly by a dinosaur. Hitchcock described a number of tracks that, like *Otozoum,* are now considered as sedimentational variants of a smaller number of basic (and actual) tracks. The species name *moodii* was named after the first discoverer of these tracks, a farmer by the name of Pliny Moodi, who in 1801 noticed them and brought them to the attention of the local newspaper. These and other tracks were later to become the main focus of studies by Professor Hitchcock. (Value range E).

Brontozoum giganteum Hitchcock (x1/5). A slab of various tracks of "birds." Edwin Hitchcock considered most of the tracks and trackways found in strata along the Connecticut River Valley to have been made by large birds. The concept of dinosaurs (which appeared in the 1840s) postdated much of his work. From E. Hitchcock, 1832. Many geology books of the nineteenth century reproduce woodcuts, like this one, from Hitchcock's works.

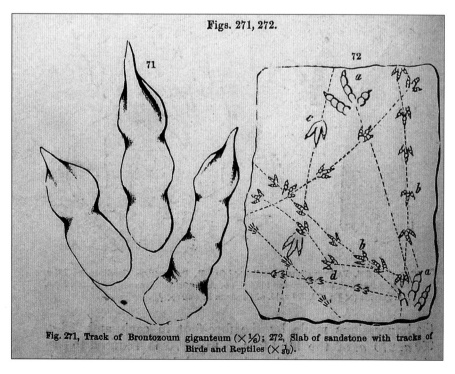

Figs. 271, 272.

Fig. 271, Track of Brontozoum giganteum (× ⅟₆); 272, Slab of sandstone with tracks of Birds and Reptiles (× ⅟₃₀).

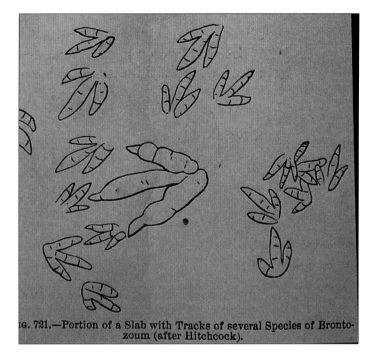

Fig. 721.—Portion of a Slab with Tracks of several Species of Bronto-zoum (after Hitchcock).

A woodcut of fossil tracks from the Connecticut River Valley from E. Hitchcock's works. A considerable variety of fossil tracks and trackways have been found in Triassic strata of eastern Massachusetts and Connecticut. Many of these have suggested close affinities to birds and indeed were first thought to be the tracks of large birds.

Awareness of extinct life forms, particularly animals, was part of eighteenth century knowledge. This illustration, from the late eighteenth century includes a dodo (middle). The illustration came from a 1795 work on natural history by Oliver Goldsmith. The Dodo was extinct by the time this illustration was made, however, much of Goldsmith's work came from earlier work by the French naturalist Comte de Buffon who wrote prior to the dodo's extinction. Awareness that life forms found as fossils represented extinct life forms was developing at this time, however, it would be a few decades later before fossil life forms would be clearly recognized as being extinct. Note the feet on these birds! The tracks they made would be like those from the Connecticut River Valley described by Edward Hitchcock. Hitchcock was familiar with Goldsmith's natural history; it was a widely disseminated work early in the nineteenth century. Hitchcock probably compared the footprints that he found with this figure in Goldsmith's "An History of the Earth and Animated Nature". (Note that in pre-1806 works an s looks like an f).

This bird is a native of the Isle of France; and the Dutch, who first discovered it there, called it in their language the *naufeous bird*, as well from its disgusting figure as from the bad taste of its flesh. However, succeeding observers contradict this first report, and assert that its flesh is good and wholesome eating. It is a silly simple bird, as may be very well supposed from its figure, and is very easily taken. Three or four dodos are enough to dine an hundred men.

Part of the text of Oliver Goldsmith's work, *An History of the Earth and Animated Nature,* discussing the Dodo. Note that an **s** looks like an **f** in pre-1806 literature.

Otozoum sp. A large number of trackways have been described from Triassic rocks of the Connecticut River Valley; some like this were made in the underlying layers of a series of silt beds, the actual footprint being found only in the uppermost layer of sediment. Tracks made in the lower layers were given specific individual names such as *Otozoum* shown here. This track, like most of the tracks and trackways found along the Connecticut River, were "coined" by Edwin Hitchcock as he was the first to investigate them. Newark Group, Upper Triassic (or Lower Jurassic), Turners Falls, Massachusetts.

A small Triassic trackway found in a parking lot excavation. Newark Group, Upper Triassic or Lower Jurassic set on the top of one of Hitchcock's books. From Hartford, Connecticut. (Value range E).

Anchisauripus tuberosus (Hitchcock). A dinosaur track from the Nash dinosaur track quarry. A commercial, fossil track and trackway quarry has been in operation for a number of years near South Hadley, Massachusetts, near where the original tracks were found in the early nineteenth century. Portland Formation, Upper Triassic, South Hadley, Massachusetts. (Value range Q).

Brochures explaining the Nash Dinosaur Track Quarry Site. This "pay dig" is close to the original site where dinosaur tracks were first discovered at the beginning of the nineteenth century. This "dig" has produced a large number of tracks over the years, which have been widely distributed in collections, both public and private.

Dinosaur track, Kayenta Formation. Dinosaur tracks are of less common occurrence in Triassic rocks of the western U.S. than they are in the eastern states, although Triassic rocks are much better exposed in the western U.S. Thick Triassic and Jurassic sandstone, like the Navajo Sandstone of Arizona, New Mexico, and Utah do have rare dinosaur tracks and trackways on infrequently occurring bedding planes. This specimen came from a highway excavation in northern Arizona where dinosaur track bearing bedding planes of the late Triassic, Kayenta Formation were found during road construction.

Eubrontes sp. (Hitchcock). A large dinosaur track originally considered by Edwin Hitchcock as the track of a very large bird. The first tracks of *Eubrontes* were found along the Connecticut River in 1803 and were described in publicity of the time as being the "Tracks of Noah's Raven." This specimen was collected from strata on the bed of the Connecticut River during a period of low water. These large dino-tracks are now considered to be Lower Jurassic in age rather than Triassic. Such tracks however, are so strongly associated with the Triassic Period that they are presented here in the Triassic section of this book. Large dinosaurs, like the one that made this track, were non-existent or rare in the Triassic Period. Newark Group, Portland Formation, Chicopee Falls, Massachusetts. (Value range C).

Collecting Triassic Fossils

As to the matter of collecting Triassic fossils, they are notably less common and available than are those from most other parts of the Phanerozoic (The Phanerozoic being that time span from the Cambrian to the Recent). This seems to hold true even at the large fossil fairs such as Tucson and MAPS expo, where fewer Triassic fossils are seen than are those from almost any other part of geologic time. The Paleozoic extinction event, even from a fossil collecting perspective, seems to have been a profound event and it took a considerable amount of time for populations to again achieve that level of biomass necessary to assure a supply of relatively abundant fossils. The author recalls, in northern Arizona,

running into piles of shale and slabby mudstone of the lower Triassic, Moenkopi Formation, which had been excavated and piled up in delayed road construction. Often large amounts of such slabby rock, excavated and piled up in this manner, even in strata which have few fossils, can be a bonanza for specimens, sometimes even rare ones. Over three miles of continuous piles of this interesting looking, slabby rock, yielded absolutely nothing in the way of fossils, not even trace fossils, which commonly show up on large slabs of siltstone, especially after being exposed to some weathering as were these slabs. The Moenkopi Formation was deposited just after the Permian extinction event, so this scenario of **no fossils** was really the one to be expected.

The author surrounded by Triassic red beds of the Chugwater Formation near Lysite, Wyoming. Bright red rock like this is characteristic of Triassic rocks worldwide. The red, highly oxidized iron of these red beds contrasts with the green of modern plants. Some astronomers, including George Darwin (son of the evolutionist Charles), suggested that the Earth, during the Triassic Period, might have resembled the planet Mars from space.

Triassic rocks worldwide are often made up of what are known as red beds, usually red sandstone, siltstone or shale. The red color comes from highly oxidized iron (ferric oxide), which may have been produced when the oxygen level of the Earth's atmosphere was higher than the 21% oxygen content of today.

Triassic red beds in southern Arizona (Monument Rocks). Red beds like these are generally barren of fossils.

Triassic shales and sandstone in Madagascar. Triassic fossils in Madagascar come from shale that is not the usual red beds typical of most Triassic rocks. *Photo courtesy of Warren Wagner*.

Marketplace in Madagascar. Fossils gathered from shale outcrops of the arid and rugged topography of this large island show up at marketplaces like this. Madagascar has produced a number of desirable fossils, many of which have shown up on the fossil market in the first decade of the twenty-first century.

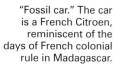

"Fossil car." The car is a French Citroen, reminiscent of the days of French colonial rule in Madagascar.

Bibliography

Dernbach, Ulrich and William D. Tidwell, 2002. *Secrets of Petrified Plants*. D'Oro Publishers, Germany.

Gould, Stephen J., 2000. "The Lying Stones of Marrakech," in *The Lying Stones of Marrakech: Penultimate Reflections in Natural History.* Harmony Books / Three Rivers Press, New York.

Hagdorn, Hans. 1999. "Triassic Muschelkalk of Central Europe," in Hess, H., W. I. Ausich, C. E. Brett, and J. Simms, *Fossil Crinoids.* Cambridge University Press.

Jahn, Melvin E. and Daniel J. Woolf, (translators), 1963. *The Lying Stones of Dr. Johann Bartholomew Adam Beringer; being his Lithograohiae Wirceburgensis*. University of California Press, Berkeley and Los Angeles.

Pinna, Giovanni, 1990. *The Illustrated Encyclopedia of Fossils.* Facts of File, New York, Oxford.

Tidwell, William D., 1998. *Common Fossil Plants of Western North America.* Smithsonian Institution Press.

Walker, Cyril and Ward, David, 1992. *Fossils, Eyewitness Handbooks. The Visual guide to more than 500 species of Fossils from around the World.* DK Publishing, Inc. New York, New York.

Chapter Three
The Jurassic Period I
—Plants to Crinoids

Land Plants

Jurassic plants consist of ferns, tree ferns, arthrophytes, cycads, and cycad-like plants (bennitailes, ginks, and various conifers).

Ferns and Tree Ferns

Here are compression fossils of fern foliage and slices through petrified tree-ferns. The tree fern *Osmundia* (or related extinct genera) were widespread plants of the Jurassic landscape. Many of these ferns were probably food for herbivorous dinosaurs.

Coniopteris sp. A variant on this small, herbaceous, and delicate fern, which resembles a fractal. Upper Jurassic, Morrison Formation, Blanding, Utah.

Coniopteris sp. A small, delicate fern from the late Jurassic Morrison Formation near Blanding, Utah. (Value range F).

A group of specimens of the fern *Coniopteris,* similar to those shown in the previous two photos. Morrison Formation, Upper Jurassic, Blanding, Utah.

Millerocaulis (dunlopi). A petrified osmunditid tree fern, trunk slice showing characteristic internal pattern. Walloon Coal Measures, Wandoan, Queensland, Australia. (Value range F).

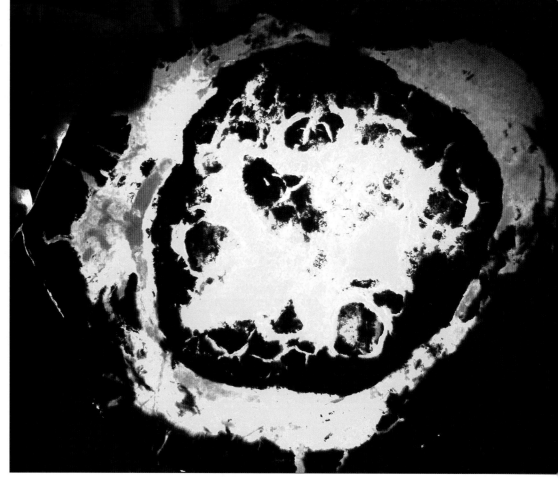

Osmundites sp.? A slice through a silicified tree fern trunk, probably that of an osmundaceous tree fern. The specimen (a sliced slab) has been backlit and is seen by transmitted light. This makes an attractive display similar to backlit agate slabs. Morrison Formation, southern Big Horn Mountains, Wyoming. (Value range F).

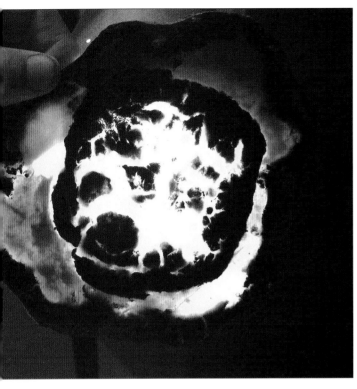

Osmundites sp. Another slice through a log of what is believed to be the trunk of a silicified tree fern. The Morrison Formation is well known for its dinosaur fauna and the foliage of these tree ferns were probably food for herbivorous dinos.

Osmunda-like fern frond. Jurassic rocks of Liaoning Province in northeast China are well known for diverse and high quality fossils. These ferns are preserved in an olive drab mudstone deposited in a set of ancient river and lake deposits of Late Jurassic and early most Cretaceous age. Yixian Formation, Yixian Province, Northeast China. (Value range F).

Osmunda-like frond. The family Osmundaceae consists of a large group of ferns and tree ferns, which were abundant during the Mesozoic and which are still abundant plants in the modern tropics. Upper Jurassic, Yixian Formation, Yixan County of Liaoning Province, northeast China. (Value range F).

Two leaves of *Osmunda*-like foliage from Yixian County, Liaoning Province, Northeast China. These plant fossils are collected "incidentally" by farmers who open pits in the fossiliferous mudstones and paper shales of which the "prize" fossils are small, feathered dinosaurs. Considerable controversy surrounds this activity; however, it is the author's opinion that it's better for fossils to be collected and distributed to interested persons and institutions than to leave them in the ground. The author has seen too many good fossil localities lost by urbanization, exurban sprawl or boorish, rich, and uncooperative landowners to feel otherwise. Liaoning Province farmers are poor and their collecting fossils for monetary gain seems, overall, to be a win-win proposition for all concerned. (Value range G)

Otozamites feistmanteli. A "cycad-like" fern preserved in a hematite concretion. Durakai, Queensland, Australia.

Cladophlebis sp. A type of early Mesozoic fern-like foliage often associated with cycads and cycad-like plants. This may be the foliage of the plants responsible for many of the petrified logs found in Triassic and Jurassic strata in the western U.S. This specimen comes from the Yorkshire coast of England, which has produced a wealth of mid-Mesozoic plants. Redcliff Rock, Cayton Bay, Yorkshire, England.

Conifers

Conifers (or what are believed to be early conifers) first appear in the late Paleozoic (*Cordaites* and *Walchia*). Fossil pollen of coniferous plants is known from the Triassic and is backed up by an abundent fossil record of coniferous wood like that of the Petrified Forest of Arizona (Chinle Formation) and the Upper Jurassic Morrison Formation of the Colorado Plateau. Coniferous foliage like that of Sequoia or *Metasequoia* is rare in Jurassic strata; some sort of cycad-like foliage may have been associated with plants represented by Triassic and Jurassic petrified logs. Various types of fern-like, compression fossils such as *Cladophlebis* sp. may also have been foliage associated with large trees represented by Jurassic petrified logs.

"Conifer?" A distinctive and peculiar fossil plant, possibly an aberrant conifer from Alaska. These are associated with cycad fronds and may represent a poorly known plant fossil from Alaska or even a plant fossil that has not been previously described. Alaska is a big and geologically complex region and its fossils remain mostly unknown. Jurassic siltstones, Talkeetna Mountains near Glen Allen, southern Alaska.

Damaeopsis sp. A early conifer possibly related to the Paleozoic plant Cordaites. These come from Upper Jurassic lake sediments (mudstones) of the Yixian Formation, a series of lake and stream deposits which have yielded a variety of fossils from pits dug into the fossiliferous beds by local farmers. Yixian County of Loaoning Province, northeast China. (Value range E).

68

Araucarioxylon sp., coniferous petrified log section. Conifers of various types and sizes were a dominant element of the Jurassic landscape. Morrison Formation, Upper Jurassic, Natrona County, southern Big Horn Mountains, Wyoming. (Value range F).

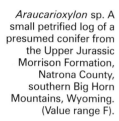

Araucarioxylon sp. A small petrified log of a presumed conifer from the Upper Jurassic Morrison Formation, Natrona County, southern Big Horn Mountains, Wyoming. (Value range F).

A complete fossil araucarian cone from volcano-clastic sediments, Cerro Caudarado, Patagonia, Argentina. Such fossil cones, until around 2000, were quite expensive and hard to come by. Complete cones such as this one are still fairly pricey and desirable, however, partial specimens, which usually exhibit good detail, have come onto the fossil market in quantity and can be acquired at relatively low prices. These attractive and unusual fossils were preserved by the original plant being quickly buried in volcanic ash. Such quick burial in high silica ash results in rapid replacement of the original woody tissue, preserving considerable detail. In some instances, entire branches containing multiple cones have been preserved in these tuffaceous sediments. The burial of limbs, cones, and logs in volcanic ash by Mt. St. Helens in 1981 represented the same scenario which took place in southern South America (Patagonia) during the Jurassic Period when a coniferous forest was quickly buried by violent volcanic eruptions. (Value range E).

Araucarian Conifers

These petrified cones from Patagonia, are some of the most striking plant fossils of the Mesozoic Era.

Araucaria sp. petrified araucarian cone. A slice through a cone from volcano-clastic sediments (tuffs) near "Cerro Cuadrado," Santa Cruz (Patagonia), Argentina. Such petrified cones are exquisitely preserved by having been buried and quickly replaced by silica in siliceous volcanic ash. (Value range E).

Araucaria sp. (Petrified pine cones). Two partial cones preserved in quartz from the Cerro Cuadrado region of Patagonia, Argentina. (Value range F, single cone).

Araucaria sp. Sliced sections of partial Patagonian fossil pine cones. Some of the cones found in the tuffs of the Cerro Cuadrado region are highly silicified (agatized) such as the specimen on the left (both halves). Complete branches, bearing cones, have been excavated from these tuffs, however such specimens, unlike these partial cones, can be quite expensive. Volcano-clastic sediments (tuff) near Cerro Cuadrado, Patagonia, Argentina. (Value range F, single pair).

Arthrophyte

Jurassic scouring rushes were not very different from those of today, except being larger.

Ephedrites chenii. This is a member of the Arthrophyta, a plant division represented today by the horsetail or scouring rush. The late Paleozoic genus *Calamites* was the largest arthrophyte to have lived. Mesozoic arthrophytes can still be fairly large, as this section from mudstones of the famous Liaoning Province fossil beds, shows. Yixian Formatin, Yixian County, Liaoning Province, northeastern China. (Value range F).

Cycads

Cycads and cycad-like fossil plants (Cycadeloides or bennitales) were among the dominant plants of the Jurassic Period. Cycad-like fossil leaf compressions are usually hard to distinguish from the leaves of true cycads unless reproductive structures are preserved (seeds or cones) with them, which they usually are not.

Nilsonia sp. A cycad-like fern from the Yixian Formation, Yixian County, Liaoning Province, northeastern China.

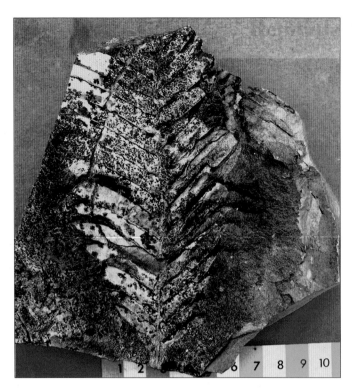

Zamites gigus. This large cycad or cycad-like frond is a widespread Jurassic fossil plant. Cycads are plants of the tropics, species living today survived the terminal Mesozoic extinction event. These large cycads from Alaska originally lived under tropical conditions. Tropical plants in Alaska at first might seem "incongruous," however many of the rocks making up this part of North America were formed at totally different parts of the globe and were later transported by tectonic forces to their present northern location. Such seems to be the case with the sediments which yield these cycads from the Talkeetna Mountains of Alaska.

Zamites gigas. A group of terminal portions of fronds from Jurassic rocks of the Talkeetna Mountains of southern Alaska. Many of the "cycads" of the Mesozoic are believed, by paleobotanists, to have belonged to an extinct plant division known as the cycadeloides or Bennitailes (or Bennettites). Reproductive structures associated with these fossil "cycads" are quite different from that of modern cycads. Reproductive structures are unknown with fossil "cycad" vegetation, so fossils like these from Alaska are unclear as to their actual taxonomic position. (Value range G, single specimen).

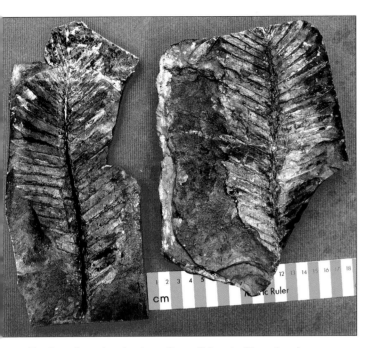

Zamites gigas. Another large "cycad" frond with part and counterpart. Upper(?) Jurassic. Talkeetna Mountains, near Glen Allen, Alaska.

Partial, single frond of *Zamites gigas*. Talkeetna Mountains, southern Alaska.

Zamites sp. A large cycad frond compression similar to the above specimens. This specimen of *Zamites*, preserved as a compression in siltstone, is very similar to the Alaskan specimens even to the rock in which it is preserved. Kimmeridgian, Saysel France. (Value range F, single specimen).

Part of a cycad "cone" from Late Jurassic strata of the Black Hills, South Dakota. A number of these large cycad cones (which resemble a group of pineapples) were found at the end of the nineteenth century but were summarily collected and now all reside in institutional collections. Specimens of these cycad "pineapples" are now quite rare and desirable as they are often preserved with colorful agate and jasper. Lakota Formation, Upper Jurassic, Black Hills, South Dakota.

A small fragment of a cycad reproductive structure, that is a "cone." Morrison Formation, Upper Jurassic, Inyan Kara, Wyoming.

A slice through a portion of a petrified cycad "pineapple." Lakota Formation, Black Hills, South Dakota.

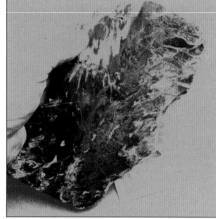

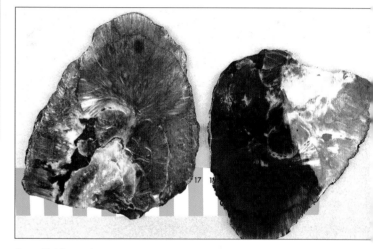

Cycad? Slices of peculiar petrified wood which may be from the lower parts of a large cycad plant. Morrison Formation, Inya Kara, Wyoming.

An early twentieth century textbook illustration of a group of the Black Hills "cycad pineapples." Nineteenth and early twentieth century geology textbooks can be a source of information on fossils and fossil "resources" which are not well known today. Such is the case with this illustration of a group of cycad "pineapples" from a 1916 textbook. The specimen illustrated was one of a number of cycad specimens collected by G. R. Wieland of Yale University. Professor Wieland was a proponent of establishing "Cycad National Monument" at the northeastern part of the Black Hills; however, most of the cycad specimens had already been collected (by him) before it was established.

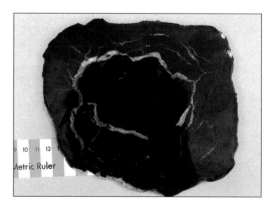

Cycad? Relatively large masses of petrified wood like this are found in the Upper Jurassic Morrison Formation and may be from cycads or they may be from conifers; both were plants common in the late Jurassic. They are often highly silicified and dark in color as seen here and thus exhibit little diagnostic structure. Morrison Formation, Upper Jurassic. Southern Big Horn Mountains, Natrona County, Wyoming.

A portion of a small silicified cycad. Such fragments turn up in late Jurassic rocks of the western states, however complete cycad "pineapples" or even large chunks of them are rare. Morrison Formation, Upper Jurassic, Inyan Kara, Wyoming. (Value range F).

Ginkos

Ginko trees, like cycads, were major types of vegetation of the Jurassic landscape. Many of the fossil logs and much of the petrified wood found in Jurassic strata may be from large ginko trees; however, it's difficult to distinguish petrified ginko wood from that of a conifer.

Ginko huttoni with a modern ginko leaf (*Ginko biloba*). The modern ginko lacks the sectioned appearance of early and mid-Mesozoic ginko leaves such as seen on this fossil. Estuarine Series, Middle Jurassic, Scarborough, England.

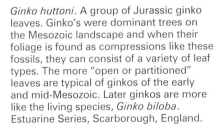

Ginko huttoni. A single leaf of a Jurassic ginko. Estuarine Series, Middle Jurassic, Scarborough, England.

Ginko huttoni. A group of Jurassic ginko leaves. Ginko's were dominant trees on the Mesozoic landscape and when their foliage is found as compressions like these fossils, they can consist of a variety of leaf types. The more "open or partitioned" leaves are typical of ginkos of the early and mid-Mesozoic. Later ginkos are more like the living species, *Ginko biloba*. Estuarine Series, Scarborough, England.

A single *Ginko huttoni* leaf. Estuarine Series, Scarborough, England.

Marine Invertebrates -Sponges and Corals

These are Jurassic fossil sponges.

Rehmmania sp. A sponge associated with late Jurassic coral reefs of southern Germany. Weissen Jura, Upper Jurassic, Heidenheim, Germany.

A group of silicified sponges found associated with coral reefs of the Late Jurassic of Germany. Malm Zeta, Weissen Jura, Heidenheim, Germany. (Value range G).

These are Jurassic fossil corals.

Montlivaltia zitteli. A group of solitary corals from Upper Jurassic limestone of southern Germany, Swabish Albs or Bavaria "Alps." Corals of the Mesozoic (and Cenozoic Eras) are known as hexicorals for the multiples of six septa. The older Paleozoic corals, known as tetracorals, are a group of coral type that went extinct at the end of the Permian and have multiples of four septa. Malm Zeta, Weissen Jura, Heidenheim, Germany, Swabish Albs of Bavaria (southern Germany).

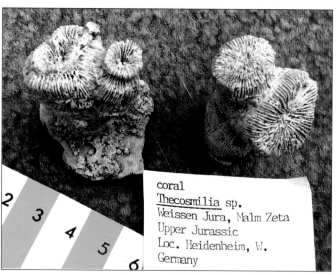

Thecosmilia sp. Two groups of this abundant Jurassic coral from coral reefs of the Swabish Albs of southern Germany. Malm Zeta, Upper Jurassic, Heidenheim, southern Germany.

Thecosmilia sp. A nice cluster of this distinctive Jurassic coral.
Heidenheim, southern Germany.

Thecosmilia sp. The four coralites shown here were "cloned" from a single corallite of this
Jurassic genus. Malm Zeta, Upper Jurassic, Heidenheim, southern Germany.

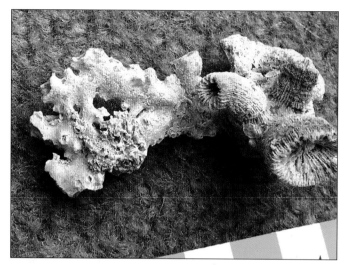

Thecosimilia sp. Attached to calcareous algae. Malm Zeta, Upper Jurassic, Heidenheim, southern Germany (Bavaria). (Value range F).

Aplosmilia sp. A silicified colonial coral from Upper Jurassic coral beds or reefs. Malm Zeta, Heidenheim, southern Germany.

Aplosmilia sp. Bottom portion of a colony of this colonial coral. Malm Zeta, Upper Jurassic, Heidenheim, southern Germany.

Cyathophora bourgueti. A colonial coral! Malm Zeta, Weissen Jura, Heidenheim, southern Germany.

Two small specimens of the colonial coral *Cyathophora* sp., Malm Zeta, Weissen Jura, Heidenhein, Germany.

Montivalcia sp., a delightful coral specimen.

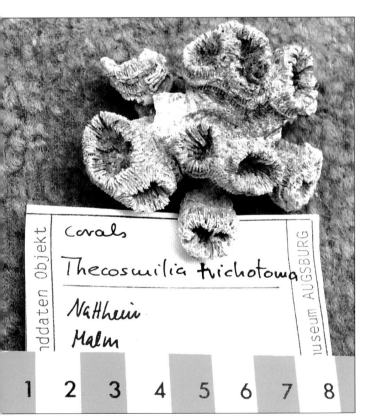

Thecosmilia trichotowa. Malm Zeta, Natthheim, southern Germany.

Large *Cyathophora* coral "head." Malm Zeta, Weissen Jura, Heidenheim, southern Germany. (Value range F).

A Fossil Jellyfish (Medusa)

It takes exceptional conditions to preserve a jellyfish as a fossil. Strata containing such extraordinary fossils can yield other exceptional fossils as well and may be, as is the Solnhofen Plattenkalk, which preserves this jellyfish, a paleontological window on geologic time.

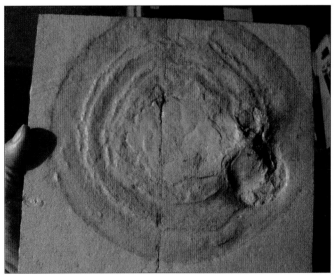

Jellyfish impression. The slabby limestone near the Bavarian town of Solnhofen have preserved a quantity of rare and desirable fossils like this impression of a jellyfish. The Solnhofen Plattenkalk comprises one of the paleontological "windows" or *lagerstatte* that represents a locality where an exceptional range of organisms have been preserved, which includes preservation of soft tissue—soft tissue such as the delicate form of this jellyfish. This fossil was made by the impression of the animal (90% water) in fine grained lime muds of a lagoon, which was protected from the open ocean and its waves by a band of coral reefs. Malm Zeta, Weissen Jura, Upper Jurassic, Solnhofen, southern Germany. (Value range D).

Brachiopods

Brachiopods are mostly associated with Paleozoic marine strata. Jurassic rocks of the Tethy's seaway can, however, locally contain them in abundance.

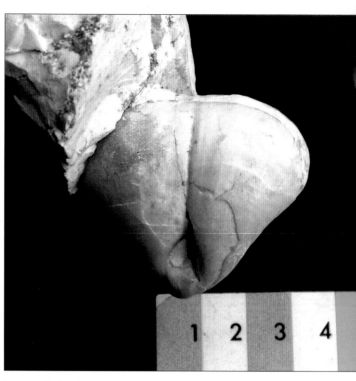

Pygope dyplia. An example of a peculiar Jurassic "keyhole" brachiopod. These brachiopods are unique to the Upper Jurassic limestone of Europe and the Arab Middle East. Titonian, Stallavna, Italy.

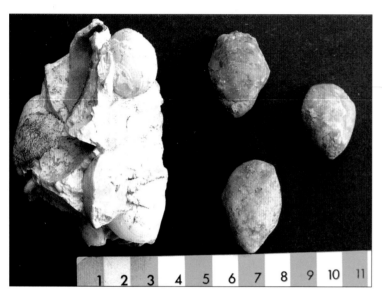

Terebratula sp. A Mesozoic brachiopod associated with Jurassic coral reefs of the Weissen Jura of Germany. Left: specimen in limestone. Right: single, small specimen. Malm Zeta, Weissen Jura, Nattheim, southern Germany.

Rhynchonella sp. A brachiopod similar to those found in the Paleozoic. These brachiopods, of the European Jurassic, are always associated with coral reefs and related white limestone deposited in very clear water. Such limestone was associated with seaways that covered a large part of southern Europe as well as part of the middle eastern Arab states such as Saudi Arabia and Iraq. (Value range G).

Terebratula sp. Another group of this locally common brachiopod from coral reef horizons of the Upper Jurassic, Malm Zeta beds of southern Germany.

Rhynchonella texebratnea. Malm Zeta, Upper Jurassic, Weissen Jura, Nattheim, southern Germany. (Value range G).

Rhynchonella sp. A group of brachiopods similar to those of the previous photo, but from southern Germany. (Value range G).

Rhynchonella sp. Late Jurassic limestone of Saudi Arabia, Iraq, and Iran have an invertebrate fauna similar to that of southern Germany which, during the Jurassic Period, was covered by the Tethy's seaway in which limestone were deposited. Partial specimens at the bottom are Paleozoic specimens of the same genus. Late Jurassic, Saudi Arabia. (Value range G).

Terebratula sp. Specimens of this common Jurassic brachiopod from Saudi Arabia. These brachiopods come from Jurassic limestone of Saudi Arabia where the same seaway of the Tethy's Ocean deposited strata similar, and of the same age, as that found in Germany. Throughout most of the Arab Middle East, Jurassic strata lie deeply buried by younger layers. Jurassic strata in the Middle East is also some of the host rock for the prolific petroleum occurrences in that part of the world. (Value range G, single specimen).

Echinoderms-Crinoids

Crinoids belonging to modern crinoid orders can be locally abundant in Jurassic rocks, particularly those of Europe.

Pentacrinus nicoletti. Crinoids in crinoidal limestone from Normandy, France. *Courtesy of Dept. of Geology, Courtney Werner Collection, Washington University*.

Chariocrinus andeae (De Sur). Thin limestone beds, almost completely made up of pieces of this delicate crinoid occur in Middle Jurassic strata of northern Switzerland. The crinoids are often compressed together and distorted as seen on this slab. Bajocian, Liestal, Canton of Basel, Switzerland. (Value range F).

Isocrinus knighti Springer. Small crinoids have been found associated with thin limestone beds of Wyoming and Utah. Pentagonal stems are particularly characteristic of Jurassic crinoids. Jurassic rocks of the Rocky Mts. and the Colorado Plateau are predominately non-marine unlike Jurassic rocks of Europe and often lack fossils. Middle Jurassic, Sundance Formation, St. George, Utah.

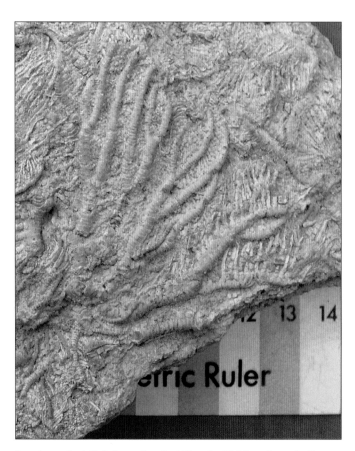

Isocrinus knighti. Two slabs of specimens of this delicate crinoid. Stems of *Isocrinus* as well as those of other Jurassic crinoids exhibit delicate extensions (cirri) which come off of the crinoid stem as seen here. Middle Jurassic, St. George, Utah. (Value range G, single specimen).

Isocrinus nicoleti. A Jurassic crinoid in crinoidal limestone similar to that from Switzerland. Bajocian, Middle Jurassic, Neufchateau, France. (Value range E).

A group of partial specimens of *Isocrinus*. Sundance Formation, Middle Jurassic, St. George, Utah.

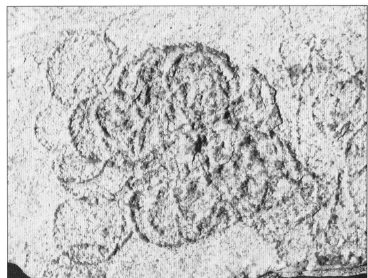

Digonriia digona. A free swimming crinoid in thin-bedded, slabby limestone of the late Jurassic. The best known occurrence of this unique type of limestone and its unique fossils is in the quarries near Solnhofen, southern Germany. This specimen, however, comes from similar limestones which occur in eastern France. (Value range F).

Close-up of the arms of the above free swimming crinoid.

Saccocoma tenella. These are small, free swimming crinoids that are found on the bedding surfaces of limestone from the quarries near Solnhofen, southern Germany (Bavaria). Extensive quarrying of this slabby, fine grained limestone for a variety of purposes, including the stone used in lithography, has produced a great number of superb fossils, which include specimens of the toothed bird *Archeopterix lithographica* and various types of pterdactyals. This fine grained limestone is capable of preserving impressions of the soft parts of animals that fell into the bottom of a lagoon where fine-grained lime mud, the source of the limestone, was deposited. Many Solnhofen fossils over the years have come onto the fossil market; these small, free swimming crinoids are the most abundant fossils in the formation. (Value range G).

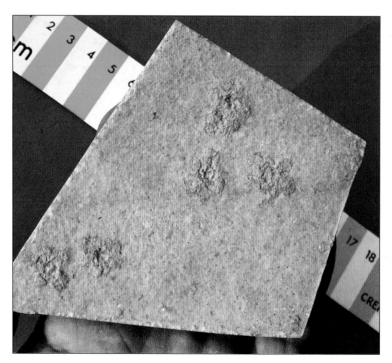

Slab of Solnhofen limestone with multiple *Saccocoma* specimens. These free swimming crinoids can be quite abundant in some slabs of Solnhofen Plattenkalk. Other than these small crinoids, however, fossils are rather rare in these layers. The actual abundance of fossils from Solnhofen is a consequence of the large amount of rock quarried and the fact that most of it gets examined either by quarry workers or masons who then set aside the fossils, which find their way to collectors and dealers. (Value range F).

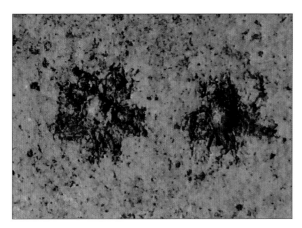

Saccocoma sp. Close up of the specimens at the left of the previous photo.

Three mangled specimens of *Saccocoma*. Specimens such as these have been suggested to have been crinoids which were partially eaten by belemnites, which were frequent visitors to the intertidal lagoons when the Solnhofen Plattenkalk was deposited.

Slab with multiple specimens of *Saccocoma*. This is the only mega-fossil which is really abundant in the Solnhofen Plattenkalk. Other fossils are actually relatively rare. (Value range F for slab).

Millericrinus milleri. A crinoid holdfast from coral bearing facies of the Weissen Jura or Upper Jurassic of southern Germany. These crinoids grew among corals in relatively agitated water that usually broke up the crinoids. Weissen Jura, Malm Zeta, Heidenheim, southern Germany.

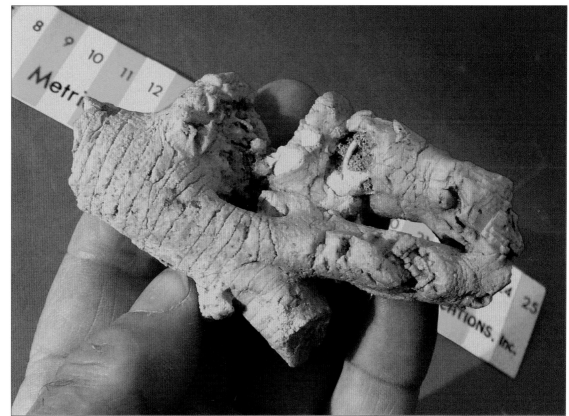

Millericrinus milleri. Two calyxes (heads) of crinoids that lived associated with corals and coral reefs of the coral reef facies in which the Weissen Jura sediments were deposited. This facies (or depositional environment) was a relatively high energy one, which generally broke up the crinoids so that complete specimens are rare. Weissen Jura, Malm Zeta, Heidenheim, southern Germany.

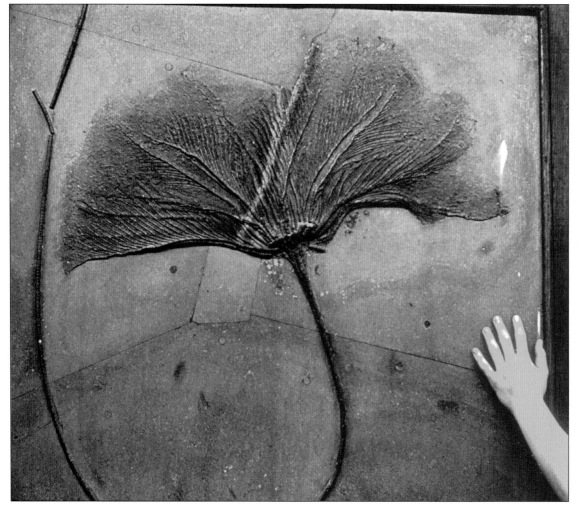

Seirocrinus subangularis A giant crinoid from the Posidenierschiefer of Holzmaden, Germany. Specimen mounted on the wall of Wilson Hall, Geology Dept., Washington University and now mounted in the Dept. of Earth and Planetary Sciences building, Washington University, St. Louis, Missouri.

Echinoderms-Echinoids

Echinoids (or sea urchins) become well established in the Jurassic. Most of these are the radially symmetrical cidarids; however, the bilateral irregular echinoids also appear for the first time.

Plegiocidaris (Cidaris) coronata (Goldfuss). A common sea urchin of the coral reef facies of the Late Jurassic (Malm Zeta) of southern Germany. Nattheim, southern Germany.

Plegiocidaris (Cidaris) coronata (Goldfuss). Two specimens of this urchin from the Weissen Jura coral reef facies. Nattheim, southern Germany. (Value range F).

Plegiocidaris (Cidaris) coronata. Three small specimens of this nice cidarid urchin. Late Jurassic, Malm Zeta, from a cement quarry near Nattheim, Germany.

Cidaris sp. Specimens of this Jurassic urchin are found in late Jurassic rocks of eastern Moracco, Algeria, and Saudi Arabian areas which were covered by the same seas which covered southern Germany. Upper Jurassic, eastern Morocco. (Value range G, single specimen).

Large, inflated, club-like spines are found on the late Mesozoic genus *Tylocidaris.* A similar genus is found in the late Jurassic of the Tethy's seaway. A *Cidaris* specimen is placed by the club-like spines to show spine placement. Malm Zeta, Nattheim, southern Germany.

Galerites sp. (Jew stone). An internal mold (steinkern) of an echinoid preserved in chert (flint) found in a brook in the Swabish Albs (Alps) of southern Germany. Such specimens have been found associated with the rocks in streams of this region for hundreds of years. During the Middle Ages fossils like the one shown here, found as a stream pebble, were observed and collected but their origin remained a mystery. They were, at that time, assigned mystical properties and referred to as "Jew stones". Weissen Jura, southern Germany. *Courtesy of Joseph Schraut collection.* (Value range F).

Clypeus ploti Salter. A Jurassic irregular echinoid. The irregular echinoids, with their bilateral rather than radial symmetry, became abundant in the Cretaceous and are the most widespread echinoid in today's oceans. *Clypeus* is one of the earliest of the irregular echinoids where they are found in the late Jurassic rocks of England. Oolite, late Jurassic (Liassic), Gloucester, southern England. (Value range F).

Hemicidaris crenularis. A cidarid echinoid from France similar to the cidarid urchins of Germany. Late Jurassic, Oxfordian, Wazicny, northern France.

Echinoderms-
Starfish and Brittle-Stars

Jurassic asteroids are desirable fossils—many come from Europe, including these from the Solnhofen Plattenkalk.

Astropecten rectus Wright. Another view of the above asteroid specimen.

Astropecten rectus Wright. A large asteroid (starfish) found in limestone rubble along the Yorkshire coast. This really nice fossil is an example of what can be found serendipitously while looking at rocks with a suitable search image. Coralline beds, Oxfordian (Liassic), Upper Jurassic, Pickering, Yorkshire, England. (Value range E).

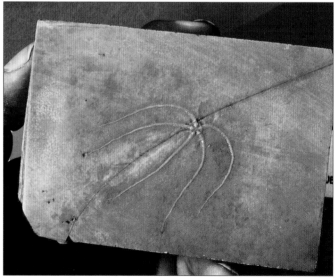

Geocoma carinata Schlotheim. A delicate brittle star from the Solnhofen Plattenkalk of Solnhofen, southern Germany. Malm Zeta, Upper Jurassic. (Value range F).

Geocoma (Ophiopinna) carinata Schlotheim. Another specimen from the Solnhofen Plattenkalk quarries. Nalm Zeta, Hienheim, southern Germany. (Value range F).

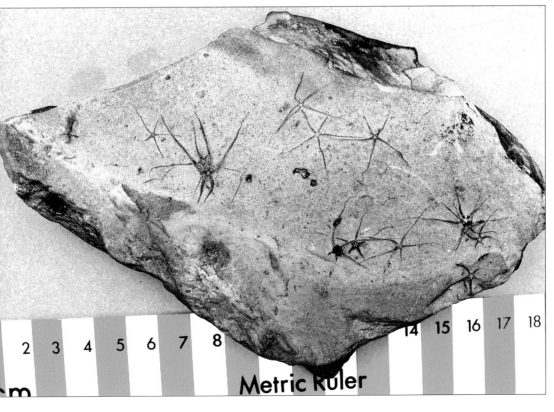

Geocoma ekegans. A delicate brittle star from late Jurassic (Callovian) strata of France. LaVoulte sur Rhone, Ardiche, France. This locality has produced many attractive and desirable brittle star fossils. (Value range E).

Bibliography

Dernbach, Ulrich and William D. Tidwell, Ed., 2002. *Secrets of Petrified Plants.* D'Oro Publishers, Germany.
Tidwell, William D., 1998. *Common Fossil Plants of Western North America.* Smithsonian Institution Press. Washington D. C. and London.

Geocoma elegans. Another delicate brittle star from LaVoulte sur Rhone, Ardiche France. Late Jurassic (Callovian). (Value range F).

Geocoma sp. A group of relatively large brittle stars from the LeVoulte sur Rhone locality of northern France. Late Jurassic. (Value range E).

Chapter Four
Jurassic II
— Bivalves to Dinosaurs

Mollusks-Bivalves (Pelecypods)

Oysters are bank-forming, specialized pelecypods that first appear in abundance in the Jurassic.

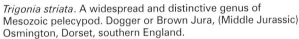

Trigonia striata. A widespread and distinctive genus of Mesozoic pelecypod. Dogger or Brown Jura, (Middle Jurassic) Osmington, Dorset, southern England.

A pelecypod from Jurassic rocks of Cuba that came through the St. Louis Academy of Science in the nineteenth century.

Metamorphosed pelecypod. A relatively rare fossil from a sequence of deep sea sediments that were "welded" to the western edge of North America during the Mesozoic Era and known as the Franciscan Series. Sediments of the Franciscan Series were originally deposited on the floor of the Pacific Ocean in the Mesozoic Era when the Pacific Ocean was smaller than it is today. Franciscan sediments were carried eastward toward North America by sea floor spreading and then "plastered" to its edge. This happened before the North American Plate overrode the mid-oceanic ridge of the Pacific. This specimen was one of a number that ended up in St. Louis in the nineteenth century through the St. Louis Academy of Science. They may have come through Joseph Le Conte who had contacts with members of the Academy in the mid-nineteenth century and who became the state geologist of California when it became a state in 1850. (Value range E, unusual fossil).

Ostrea sp. These are early fossil oysters. Oysters, which are specialized pelecypods, appeared in the Triassic and became locally abundant in the Jurassic Period. They can occur in large numbers where they formed "oyster banks" similar to the manner in which oysters live today. Oxfordian, Upper Jurassic, Normandy, France.

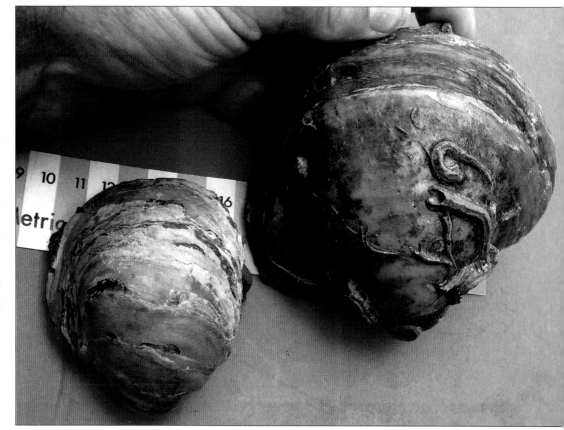

Ostrea sp. Two fossil oysters with a calcareous worm tube formed on the right hand specimen. Oysters can be common fossils in shallow water marls of the late Mesozoic. The oyster shell "environment" offered an ecological niche which was exploited by marine worms like *Serpula* sp., the marine worm which secreted the curved tube on this specimen. Serpula becomes a common fossil itself in the late Jurassic and later where it attaches itself to hard surfaces like the shell of this oyster. Upper Jurassic, Normandy, France. (Value range F).

Tetragonlepis sermicinctus. A large Jurassic oyster from coral reef horizons of the late Jurassic, Lias Episelon, Natthein, Germany.

Ostrea cristagalli. Oxfordian, (Upper Jurassic), Normandy, France.

Ostrea cristagalli. Oxfordian, Willers-s-Mer, Normandy, France.

Gryphaea arcuata. Weissen Jura, Lias alpha-1. Deizisau, Wurttemburg, Germany.

Gryphaea sp. The genus *Gryphaea* is a widespread coiled fossil, a "coiled oyster" of the Cretaceous Period that goes extinct at the end of the Cretaceous. These fossils represent one of the earliest appearances of *Gryphaea.* Upper Jurassic, Dorset, southern England.

Mollusks-Gastropods

Gastropods (snails) are mollusks that have left
a large and complex fossil record.

A high, spired gastropod from the late
Jurassic, southern Germany, Bavarian
coral reef facies, Weissen Jura, Malm Zeta,
Heidenheim, Germany.

Pleurotomaria granulata. Another species of *Pleurotomaria* from
coral reef facies of the Late Jurassic, Malm Zeta, Heidenheim,
Germany. (Value range G).

Pleurotomaria sp. *Pleurotomaria* is a long ranging gastropod which
is still living in the Pacific Ocean today, large shells of it being prized
by shell collectors. The specimen to the left has the original shell,
that on the right is an internal mold of this gastropod. Weissen Jura,
Heidenheim, Germany.

Mollusks-
Cephalopods-Nautaloids

Nautaloid cephalopods are the only shelled cephalopods living in today's oceans. They are attractive fossils, particularly when the chambered part of the shell, the phragmacone, is well preserved.

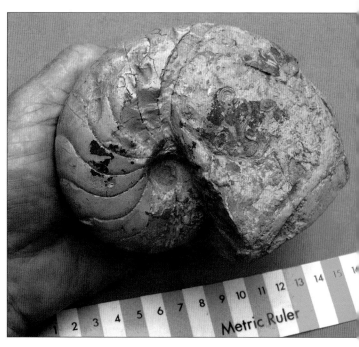

Cenoceras (*Metacenoceras*) A smaller specimen of the same nautiloid as shown above from the same molluscan rich, Jurassic strata. Weissen Jura, Wurttemburg, southern Germany.

Cenoceras (*Metacenoceras*) sp. A coiled nautiloid cephalopod from Late Jurassic strata. This genus is quite close to the living Nautilus which is the only living genus of a shelled cephalopod. The shell on this specimen has been partially removed to show the slightly curved chamber walls or sutures. Nautiloids differ from ammonoids in their possession of simple, usually non-convoluted septa which on ammonoids take on a complex configuration. Weissen Jura, Wurttemburg, Germany. (Value range F).

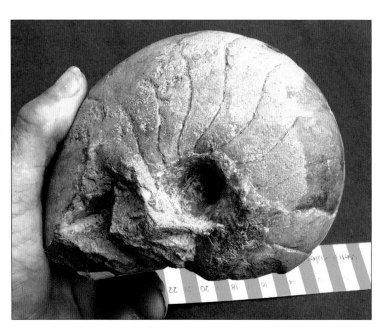

Cenoceras sp. Nautaloid (side view). The same specimen as in the previous picture which shows the nautiloid suture pattern nicely. Upper Jurassic, Weissen Jura.

Cenoceras (*Metacenoceras*) sp. A highly polished nautilus-like cephalopod worked out from Middle Jurassic oolitic limestone of the Atlas Mountains of Morocco. These fossils showed up at the Tucson Show in 2000. Identical nautiloids are found in Middle Jurassic, oolitic limestone of England and France. (Value range F).

Mollusks-
Cephalopods-Ammonites

Ammonites are a huge group (subclass) of cephalopods which reached their zenith of diversity during the Jurassic Period. Ammonites have intrigued rock conscious persons of Europe and elsewhere since the Middle Ages when they were given religious significance; these fossils have "tweaked" many a curious mind over the centuries!

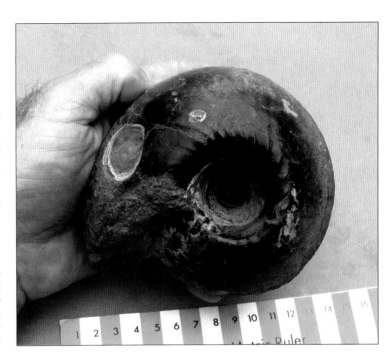

Cadoceras sublaeve. This is an ammonite which superficially resembles a nautiloid. The presence of ornamentation and ammonite sutures shows it to be an ammonite. Many of these ammonites from England have come from old collections, some dating back to the early nineteenth century. Callovian, Kellaway's Rock, Wiltshire, England. (Value range F).

Kellaway's Rock outcrops have produced a number of fine ammonites over two centuries such as this specimen. This is a different view of the same specimen from Kellaway's Rock as shown previously.

A discussion on Kellaway (Kellaway's Rock), (the source of previously shown ammonites) from C. Lyell, *Elements of Geology.* The illustration shows a belemnite to the right which came from a railroad cutting made in 1840 and which was a source of many nice Jurassic fossils illustrated in Victorian geologic literature.

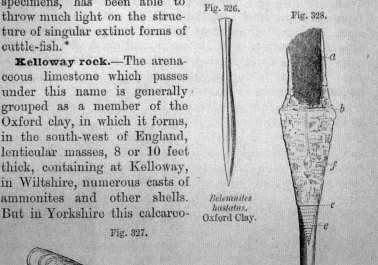

from the shells of some belemnites discovered by Dr. Mantell, in the same clay (see fig. 328), who, by the aid of this and other specimens, has been able to throw much light on the structure of singular extinct forms of cuttle-fish. *

Kelloway rock.—The arenaceous limestone which passes under this name is generally grouped as a member of the Oxford clay, in which it forms, in the south-west of England, lenticular masses, 8 or 10 feet thick, containing at Kelloway, in Wiltshire, numerous casts of ammonites and other shells. But in Yorkshire this calcareo-

Fig. 326.

Fig. 328.

Belemnites hastatus. Oxford Clay.

Fig. 327.

Cadoceras sublaeve (Sowerby). An
involute ammonite from Kellaway's
Rock. Chippenham, Whltshire, England.

A large ammonite from the
late Jurassic of southern
Germany. Malm Zeta, Weissen
Jura, Heidenheim, Germany.

Cadoceras sublaeve
(Sowerby). Another view of
the same involute specimen
shown previously.

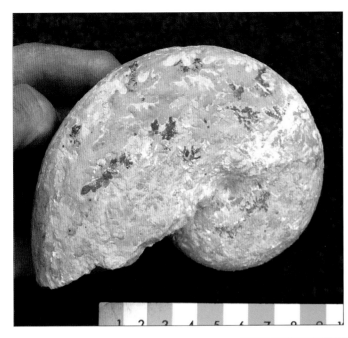

Phylloceras heterophyllum. An involute ammonite of similar appearance to a Triassic ammonoid. Most Triassic ammonoids are ceratites, this is a true ammonite that resembles a ceratite. Toarcian Stage, Lower Jurassic, Cingoli, central Italy. (Value range G).

"Inflated looking" ammonites which resemble goniatites. These Middle Jurassic ammonites have been worked out of ammonite bearing chunks of oolitic limestone which occur in the Atlas Mountains of Morocco. Such ammonites have been confused by fossil dealers in Moroccan fossils with the much older Lower Devonian ammonoids (Goniatites), which come from the same region. Arbaza, Mid-Atlas Mountains, Morocco.

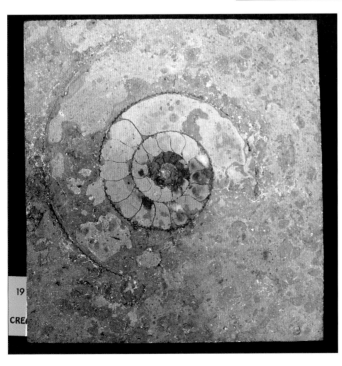

A stone floor-tile containing the cross section of an ammonite from Eichstadt, southern Germany. "Marble" (cut and polished limestone) from strata of the Weissen Jura is sliced and polished and is widely used in Germany as floor tiles. Some of these tiles, like this one, exhibit ammonite cross sections which can be seen intermixed on marble floors with other non-fossil bearing tiles; they make an interesting floor widely appreciated in Europe.

Single specimen of goniatite-like ammonite from Arbaza, Atlas Mountains, Morocco.

Liparoceras cheltiense (Murchison). These ornate ammonites came from an excavation that produced a number of them in the 1980s. Lower Lias, Blockley, Gloucestershire, southern England. (Value range F, single specimen).

Broken specimen of the above goniatite-like ammonite showing calcite crystals which line the chambers of these ammonites from the Atlas Mountains. Arbaza, Morocco.

Liparoceras cheltiese (Murchison). Same specimen as that to the right in the previous photo.

Acanthopleuroceras valdani (D'Orbigny). A nice Middle Jurassic ammonite which came through Geological Enterprises of Ardmore Oklahoma. Pliensbachian, Lower Jurassic. Laize la Villa, Cavados, Normandy, France.

Pleuroceras sp. A common ammonite in the Middle Jurassic (Lias Delta) of Germany and France. Calvados, Normandy, France.

Choffatia subbackeriae (D'Orbigny) Middle Jurassic,
Callovian, Mamers, Sarthe, France.

Pleuroceras sp. Middle Jurassic, Lias Delta, Calvados, France.

Harpoceras falcifer. Upper Lias, Lower Jurassic, Barford, St. John, England. (Value range F).

Leioceras sp. A number of these attractive, polished ammonites from Morocco appeared at the 2008 Tucson show. They are often listed by fossil dealers as the genus Shloenbacchia and as being of Lower Cretaceous age. These Middle Jurassic (Bajocian) involute ammonites are worked out of blocks of ammonite bearing, oolitic limestone from the Atlas Mountains. They show distinct vugs (or cavities) in the ammonite chambers which are lined with calcite crystals. These ammonites, which are highly polished to show the sutures, are quite attractive. (Value range E).

Harpoceras sp. An envolute ammonite from Bajocian age (Middle Jurassic) oolitic limestone of the Atlas Mountains of Morocco. Surface ornamentation on these ammonites, which usually is subdued anyway, has been removed in the process of extracting these fossils from chunks of ammonite bearing limestone. This method of extracting them by grinding is the only method capable of exposing them. With this method of preparation, the calcite filled chambers make a beautiful fossil that is as much a mineral specimen as it is a fossil. (Value range E).

Leioceras (*Shloenbacchia*) sp. A group of highly polished ammonites from Middle Jurassic, oolitic limestone of Morocco. These ammonites are similar to those which come from Jurassic oolitic limestone of southern England. They all lived in the Tethy's Ocean where their shells could drift long distances to give very cosmopolitan ammonite faunas. (Value range F, single specimen).

Dactylioceras sp. Schwartzen Jura, Lower Jurassic. One of the most common and desirable fossils from the celebrated Holzmaden "slates" of Wurttemberg, Germany. Specimens of this ammonite, severely compressed in black shale, come from quarries near the town of Holzmaden in the province of Wurttemberg, near Stuttgart, southern Germany. Fossils in the black, carbonaceous and/or petroliferous shale have been mined at Holzmaden for over 200 years where it is used for the "slate" of billiard and pool tables, as well as being mined for its large and spectacular fossils, which include complete Ichyosaurs, Plesiosaurs, and huge crinoid slabs. The ammonites from Holzmaden are always flattened on bedding planes of this hard, black shale. (Value range F).

Harpoceras (Lytoceras) falcifer. A severely compressed ammonite from the black shales ("slates") of Holzmaden, Wurttemburg, Germany. This is the second most abundant ammonite from Holzmaden after *Dactylioceras*. (Value range F).

Asteroceras obtusum. These ammonites come from the famous sea cliff exposures of Lyme Regis on the Dorset coast of southern England. Lyme Regis (or "Lyme") has been a favored locality for fossil collecting since the eighteenth century. When geology became a "new" science in the early nineteenth century, British gentlemen (and ladies too) often took up the fashionable hobby of fossil collecting. The sea cliffs at Lyme were rich in excellent fossils such as these, which could be extracted from the rocks with relative ease. Fossil collecting at "Lyme" made a cameo appearance in the 1980s movie *The French Lieutenant's Women*. Fossils are not so easily acquired today as the sea cliffs, which used to erode extensively during winter storms, have now been buttressed by large cement berms. Lyme Regis ammonites, however, like these from an old collection, turn up when they have been handed down over the years through the activities of many collectors. (Value range F, single specimen).

Dactylioceras (Tenuidactylites) tenuicostatum (Young and Bird). These ammonites were often collected in black limestone nodules or limestone lenses. Sea cliffs and low tide outcrops at Whitby, on the east coast of England about 200 miles north of London, have yielded these ammonites for hundreds of years. In the early and mid-nineteenth century, when fossil collecting became a gentlemen's hobby, thousands of quality specimens were collected that went into both private and institutional collections and now show up with some frequency as these old collections are recycled, sometimes showing up in jumble shops in London as well as in antiquarian shops. As fossil specimens have been widely traded across the Atlantic, specimens from Whitby also show up in the states, sometimes in antique shops and even at garage sales. The area still produces specimens, however not in the quantities of years past. Upper Lias-*Tenuicostatum* zone, Lower Jurassic, Whitby, Yorkshire (Yorks), England. (Value range F).

Dactylioceras tenulcostatum. The town of Whitby's coat of arms displays ammonites with snakes' heads on them. Specimens of *Dactylioceras* with carved snakes' heads have been sold to seaside tourists since the eighteenth century. These ammonites, in their black "coffins," have been known since the Middle Ages, where legend has it that St. Hilda removed all of the snakes from the region by petrifying them, then placing them in the little "stone coffins" (concretions) found along the sea shore. Whitby and another Jurassic fossil producing sea shore and its adjacent cliffs, Lyme Regis to the south, have been popular fossil producing, sea-side regions for over 200 years. Whitby, Yorkshire (Yorks), England. (Value range E).

Dactylioceras tenulcostatum. This relatively large specimen of *Dactylioceras* from Whitby came from an old collection. (Value range E).

Back of the previous specimen showing attached label from a collection made in the nineteenth century. Many English and other European ammonites from old collections retain labels like this, sometimes in a distinctive script or calligraphy.

Group of *Dactylioceras* in nodules (stone coffins) from Whitby, England, which were originally the "snakes" which St. Hilda turned into stone.

Harpoceras (Hildoceras) bifrons. Lias (Lower Jurassic). One of the other ammonites frequently found at Whitby. The genus *Hildoceras* sp., originally described, has reference to St. Hilda who in the Middle Ages was supposed to have turned all of the snakes found above the sea cliffs near Whitby into stone (ammonites) and then put them into little stone coffins. Lias (Lower Jurassic), Whitby, Yorkshire, England. (Value range E).

Dactylioceras commune. Brown Jura (Middle Jurassic), Wurttemberg, Germany. A large number of ammonite assemblages from this locality came onto the fossil market in the 1970s. They were first found during excavations of a housing development by collectors (of which there are a number in Germany) and hence salvaged from an otherwise secondary burial. (Value range F).

Dactylioceras (Perisphinctes) sp. These ammonites are almost identical to those found in Germany, France, and England, but these came from half a world away and come from Jurassic strata of the island of Madagascar. Ammonite faunas of the same age, can be very similar or identical worldwide as ammonites were swimmers that could travel considerable distances. Empty shells of dead animals also could float hundreds of miles in large groups before falling to the sea floor or washing onto a Jurassic beach. The thousands upon thousands of these ammonite shells on Jurassic beaches must have been a curious sight at the time. These Madagascar ammonites, found recently and distributed worldwide through the commercial fossil "industry," are quite attractive. The genus *Perisphinctes* and *Dactylioceras* are quite similar, with *Perisphinctes* expanding more rapidly than does *Dactylioceras*. (Value range G, single specimen).

Perispinctes sp. A large number of these ammonites, from the late Jurassic, have come from limestone around the south German town of Bopfingen. They are in a homogeneous, tan colored limestone, preserved as internal molds with no original shell material. Weissen Jura, Malm Zeta, Bopfingen, Wurttemberg, Germany. (Value range F, single specimen).

Dactylioceras sp. Another group of ammonites in tan limestone from the late Jurassic of Madagascar. Madagascar ammonites came onto the fossil market in quantity in 2006. They are similar to the ammonites from Europe, which are historically interesting in their connection with geology, fossil collecting, and science. Madagascar ammonites are currently priced considerably lower than are equivalent specimens from Europe with their historical significance. Upper Jurassic, Tulear Province, Madagascar. (Value range G, individual specimen).

Perisphinctus cf. *P. subleles.* A small specimen of this common ammonite genus which preserves original shell nacre in "gold." A number of these (golden) ammonites, with original shell nacre, have been said to be preserved with pyrite (fools gold) which they are **not**, they are ammonites whose original shell nacre has been stained with iron oxide. These were distributed by the German fossil dealer F. Krantz in the early part of the twentieth century. This specimen (and many others) was acquired in the 1920s by Professor Courtney Werner of Washington University in St. Louis.

Cycloceras nangenesti. This ammonite has been replaced with pyrite or the related mineral marcasite. A number of these come from Lyme Regis, Dorsetshire, England, where they occur in a black, organic rich layer of coal-like material. This specimen came from an old collection containing many Lyme ammonites. (Value range F).

Parkinsonia parkinsoni. A common Middle Jurassic ammonite. Lias epsilon (Braun Jura), Dogger. A number of nice Middle Jurassic fossils have been collected and preserved by German collectors from quarries in the vicinity of Sengenthal, Oberpfalz, Bavaria. Both the genus and species of this ammonite are named after Dr. Ellison Parkinson who, as an early nineteenth century medical doctor, contributed to Paleontology, but is best known for recognizing what is known as the shaking palsy or Parkinson's Disease. (Value range F).

Oxycerites sp. An involute ammonite from the Lias epsilon (Dogger), Sengenthal, Bavaria. (Value range F).

Pleuroceras sp. A partially pyritized ammonite from the Lower Jurassic of southern Germany. (Value range F).

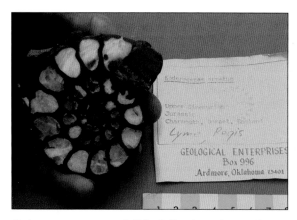

Eoderoceras armatum. Polished slice through a pyritized ammonite. The chambers of the phragmacone have been filled with calcite and a fossil like this can be looked upon as both a mineral as well as a fossil specimen. Ammonite from the Lower Jurassic rocks of Lyme Regis, Dorset, southern England. (Value range E).

Lyoceras oppalenium. Braun (Brown) Jura-alpha zone. A number of these specimens came from an autobahn cut between Holtzmaden and Eidelburg, Baden-Wurttemburg, southern Germany. (Value range F).

from the shells of some belemnites discovered by Dr. Mantell, in the same clay (see fig. 328), who, by the aid of this and other specimens, has been able to throw much light on the structure of singular extinct forms of cuttle-fish.*

Kelloway rock.—The arenaceous limestone which passes under this name is generally grouped as a member of the Oxford clay, in which it forms, in the south-west of England, lenticular masses, 8 or 10 feet thick, containing at Kelloway, in Wiltshire, numerous casts of ammonites and other shells. But in Yorkshire this calcareo-

Fig. 326.

Fig. 328.

Fig. 327.

Belemnites hastatus. Oxford Clay.

Belemnites Puzosianus, d'Orb. *B. Owenii,* Pierce.

Oxford Clay, Christian Malford.

a. Section of the shell projecting from the phragmacone.
b–c. External covering to the ink-bag and phragmacone.
c, d. Osselet, or that portion commonly called the belemnite.
e. Conical chambered body called the phragmacone.
f. Position of ink-bag beneath the shelly covering.

Ammonites Jason, Reinecke. (Syn. *A. Elizabethæ,* Pratt.) Oxford Clay, Christian Malford, Wiltshire.

arenaceous formation thickens to about 30 feet, and constitutes the lower part of the Middle Oolite, extending inland from Scarborough in a southerly direction. The number of mollusca

Ammonites Jason. One of the unusual ammonites and a belemnite that came from railway cuttings in the Oxford Clay when the British rail system was first being built. Illustration and text from Charles Lyell, *Principles of Geology*, 1835, John Murray, Publisher.

Ancyloceras calloviense. Ammonite from the Oxford Clay, Lower Jurassic. A locality southwest of London in the nineteenth century yielded a number of these compressed ammonites from a clay outcrop; this was exposed in railway cuttings, which were highly collected at the time. Old collections of English fossils can include specimens from this locality whose best known ammonite is the unusual "*Ammonites Jason.*" Oxford Clay, Christian Malford, Wiltshire, England. (Value range E).

Quenstedoceras sp. Late Jurassic of the craton of Russia. Popilany, Soviet Union. Few fossils or other geologic specimens came out of Russia during the time of the Soviet government. After its "fall," a number of specimens from various parts of Russia entered the fossil market, presumably one of the advantages of capitalism! (Value range F).

Quenstedoceras sp. These ammonites occur in a zone where their shells probably washed onto a beach in great number. Ammonites, being swimming organisms, whose shells – when the animal died – continued to float and concentrate in localized areas. Such fossils, before 1990, were rarely seen as the Soviet government prohibited export of geologic specimens, which included fossils. Popilany, Russia.

Randiceras sp. A pyritized ammonite which has had its outer shell removed to show chambers and sutures covered with a fine pyrite druse. A number of these came onto the fossil market in 2006 from the Saratov Region, Volga River, Russia. With their chambers and chamber walls covered with a druse of small pyrite crystals, these are as much a mineral as a fossil specimen. (Value range F).

A Jurassic ammonite from Gulf Coast sediments which, other than in Mexico, do not outcrop on the surface to yield fossils. Real de Catorre, near Monterey, Mexico.

Perisphinctus sp. Jurassic rocks underlie Cretaceous strata in the Gulf Coastal Plain of the southern U.S. but are not exposed, as they are always covered by younger strata. Only in the vicinity of Monterey, Mexico, has uplift occurred that brought up these Jurassic beds, the source of some excellent ammonites. Real de Catorre, Monterey, Mexico. (Value range F).

Stephanoceras sp. Two Jurassic ammonites from India. The author has seen few fossils coming from India, however a number of fine mineral specimens have recently come from this country on the geo-collectors market. (Value range F, single specimen).

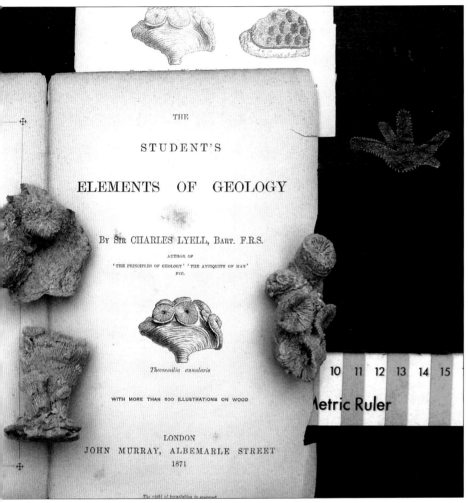

THE

STUDENT'S

ELEMENTS OF GEOLOGY

By Sir CHARLES LYELL, Bart. F.R.S.

AUTHOR OF
'THE PRINCIPLES OF GEOLOGY' 'THE ANTIQUITY OF MAN'
ETC.

Thecosmilia annularis

WITH MORE THAN 600 ILLUSTRATIONS ON WOOD

LONDON
JOHN MURRAY, ALBEMARLE STREET
1871

The title page of Charles Lyell's *Elements of Geology*. The Jurassic coral *Thecosimilia* is shown along with specimens of this same coral from southern Germany. Charles Lyell is considered as the father (or one of the founders) of modern geology, which emphasized the progression of life, as represented by the fossil record—a record preserved in rock strata, which was found to span vast amounts of time.

"Ammonites" humphresianus. Woodcut illustration from Lyell's *Elements of Geology* with a specimen of a related ammonite. The genus *ammonites* was placed into synonymy at the end of the nineteenth century and is no longer a valid ammonite genus.

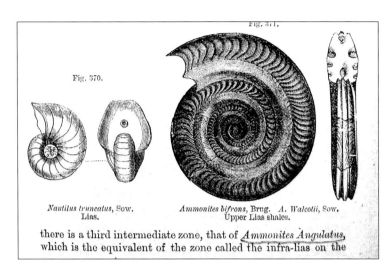

"Ammonites" bifrons. A woodcut illustration of this ammonite now known as *Harpoceras bifrons,* from the Lower Jurassic (Lias) of Whitby, England. A coiled nautiloid, similar to those found in Madagascar, is illustrated to the left. Charles Lyell was well aware of the Whitby ammonites where he collected specimens, sometimes with his close friend Charles Darwin. Mr. Darwin did venture to a few of the classic English fossil localities like Lyme, if they were not too far from his home outside of London. Charles Darwin compared fossil collecting with hunting, an activity which he engaged in as a young man, but later in life gave up as he could no longer kill an animal. The finding of a nice fossil, however, according to Darwin, gave him "the same thrill as did the hunt but without the carnage."

Ammonites from Solnhofen. Ammonites surprisingly are not common in the slabby rock of this famous locality (Plattenkalk) but occur abundantly in overlying and underlying strata. The ammonite impression at the left is in the slabby Plattenkalk; note the nice manganese dendrites to the right. The ammonite on the right comes from strata overlying the Plattenkalk, which is quarried in a nearby cement quarry. Ammonites are common here but they are not common in the slabby Plattenkalk, so famous for its unique fossils. (Value range F each specimen).

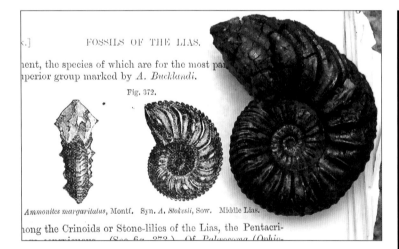

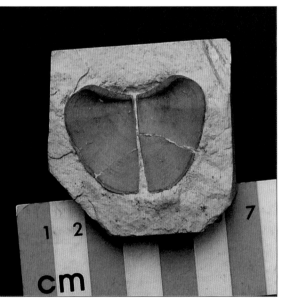

"Ammonites" bucklandi. Another ammonite illustration from Lyell's *Elements of Geology*. The species is named after William Buckland, a theologian turned geologist who was a contemporary of Charles Lyell. The Buckland-Lyell-Darwin connection is well represented in James Burke's 1990s TV series "The Day the Universe Changed," in an episode named "Darwin's Revolution." Check it out!

The calcified apertural covering or apticus from a Jurassic ammonite. These ammonite apertural covers are uncommon fossils, presumably most ammonites lacked them or if they had them they were not mineralized, so they normally would not be preserved as fossils. Solnhofen, Bavaria.

Mollusks-
Cephalopods-Belemnites

Belemnites are extinct cephalopods, distantly related to the living squid or cuttlefish. They, like ammonites, reached their zenith of diversity in the Jurassic Period.

Megatheutis giganteus. A large, Middle Jurassic belemnite. Belemnites were squid-like animals that went extinct—along with ammonites (and many other organisms), at the end of the Mesozoic Era with the Mesozoic extinction event. These are some of the largest belemnites known. Dogger delta, Middle Jurassic, Segenthal, Bavaria. (Value range E).

Salpingoteuthis sp. These are some of the most common fossils, next to ammonites, found in the black, slaty shales of Holzmaden, Wurttenberg, southern Germany; they are quite slender belemnites. (Value range F, single specimen).

Blemnopsis hostatus. These belemnites are found in compact late Jurassic limestone quarried near Eichstadt, not far from Solnhofen, Bavaria. They are often sold as curios and sometimes are made into pen set bases with the fossil belemnite resembling a ball point pen (Kugelschieffer). Weissen Jura, Malm Zeta, Eichstadt, southern Germany. (Value range F).

Cylindroteuthis (Belemnites) puzosiana. A recently collected specimen from the Late Jurassic Oxford Clay of southern England. This belemnite chambered phragmacone is to the right of the specimen. Oxford Clay, Newton Longville, England (Value range G).

Belemnites cf. *B. hastatus*. Typical English belemnites of a type characteristic of the late Jurassic of western Europe. These specimens are similar to those found in the Oxford Clay southwest of London in the nineteenth century, were they were associated with strata encountered in the "cuttings" for Briton's railway system (Value range F for group).

Pachyteuthis sp. A medium-sized belemnite from Middle Jurassic rocks of Wyoming. Large numbers of these belemnites have been collected and sold to tourists passing through central Wyoming on their way to Yellowstone or to the Black Hills. Middle Jurassic, Sundance Formation, Shell, Wyoming.

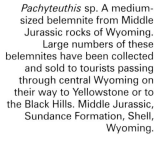

Probably no Jurassic strata is better known for its fossils than is the Solnhofen Plattenkalk of Bavaria. It is considered as a "paleontologic window." Here, numerous quarries, working thin bedded or slabby limestone deposited in a protected lagoon, has produced a wealth of fine fossils, including the toothed bird *Archeopterix*, complete small dinosaurs and pterodactyls.

A fossil shop in southern Germany specializing in German Jurassic fossils, especially those from nearby Solnhofen. The rectangular object on the doorway at the right is a cigarette dispenser; these dispensers were (and are) found lurking in unlikely places throughout Germany.

Solenhofen stone.—The celebrated lithographic stone of Solenhofen in Bavaria, appears to be of intermediate age between the Kimmeridge clay and the Coral Rag, presently to be described. It affords a remarkable example of the variety of fossils which may be preserved under favourable circumstances, and what delicate impressions of the tender parts of certain animals and plants may be retained where the sediment is of extreme fineness. Although the number of testacea in this slate is small, and the plants few, and those all marine, Count Münster had determined no less than 237 species of fossils when I saw his collection in 1833; and among them no less than seven *species* of flying reptiles or pterodactyls (see fig. 320), six saurians, three tortoises, sixty species of fish, forty-six of crustacea, and twenty-six of insects. These insects, among which is a libellula, or dragon-fly, must have been blown out to sea, probably from the same land to which the Pterodactyls, and other contemporaneous air-breathers, resorted.

In the same slate of Solenhofen a fine example was met with in 1862 of the skeleton of a bird almost entire, and retaining

Fig. 320.

Skeleton of *Pterodactylus crassirostris*. Oolite of Pappenheim, near Solenhofen.

a. This bone, consisting of four joints, is part of the fifth or outermost digit elongated, as in bats, for the support of a wing.

Explanation discussing the quality fossils found in the quarries of Solnhofen, Bavaria, by Sir Charles Lyell in his *Elements of Geology*, John Murray, publisher.

One of numerous quarries working the slabby rock of the Solnhofen Plattenkalk; picture taken in 1993.

Arthropods

The Solnhofen Plattenkalk is world renown for its fossil arthropods, which includes a variety of crustaceans.

Mesolimulus walchi. A specimen of a horseshoe crab showing pronounced spikes coming off of the animals thorax. This species is primarily known from Late Jurassic beds of Solnhofen, Bavaria. Fossil horseshoe crabs are generally rare fossils; these examples from Solnhofen are particularly nice. (Value range D).

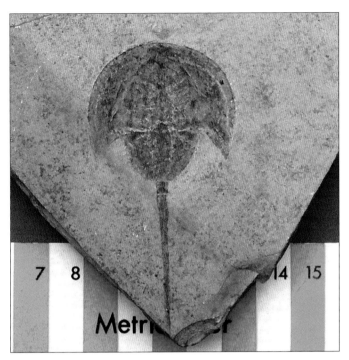

Pinnae sp. A small shrimp that is a fairly common fossil in the Solnhofen Plattenkalk. (Value range F).

Mesolimulus sp. Another specimen of horseshoe crab from the Upper Jurassic Solnhofen Plattenkalk, which lacks (were not preserved?) the thoraxial spikes. Horseshoe crabs are an ancient family of arthropods that, because of their fragile carapace, are relatively rare fossils. They are still living today. (Value range D).

Eryma leptodectylina. A fairly common shrimp of the Solnhofen Plattenkalk. Fossils are actually relatively rare in the Solnhofen beds. The large amount of rock that is quarried and the fact that most of what is quarried is examined by either quarrymen or stone masons who spot the fossils which do occur and set them aside for collectors to purchase leads to a large fossil output. Fossils are a profitable by-product of the Solnhofen quarrying operations and their sale to collectors has encouraged their preservation. (Value range E).

Pinnae speciosus. A type of shrimp that is one of the more common fossils of the Solnhofen limestone.

Aeger tipularis. This is a fossil lobster from Solnhofen, Bavaria. It has original caparace material preserved on it, a feature which varies from specimen to specimen with Solnhofen fossils.

Eryma modestiformis. A Solnhofen lobster. Crustaceans, of which shrimp and lobsters are members, become common marine arthropods after the extinction of the trilobites. Shrimp, however, are known from the late Paleozoic, where they are not generally associated with trilobites. (Value range E).

Mecochirus longimanus. Some of the finest Solnhofen fossils have original material on them, such as the carapace of arthropods or the original bone of vertebrates. Often, however, the fossil occurs as an impression which lacks original material, such as in this specimen. Such specimens are sometimes highlighted by stain or paint, a technique which makes the relatively rare fossils from the quarries stand out more clearly and produces a greater market for them. Such highlighting, if done skillfully, does not degrade the value of the fossil and makes it more attractive and easier to see. This specimen has not been highlighted and has no original carapace material; it's also hard to see. The specimen in the previous photo has been highlighted. (Value range E).

Mecochirus longimanus. Another specimen of this long armed "Krebs" (crab) from Solnhofen. This specimen has been outlined with paint, a technique used to enhance specimens lacking original material; such enhancement makes the fossil more obvious and hence more saleable. (Value range F).

Mecochirus longimanus. The long-armed lobster of Solnhofen. Besides shrimp, this is one of the more common fossil arthropods from Solnhofen. This lobster has particularly long anterior (front) appendages. Solnhofen Plattenkalk, Weissen Jura, Malm Zeta. (Value range E).

Mecochirus longimanus. Solnhofen fossils are often found on the bedding planes of the Plattenkalk; however, with some specimens, the fossil is embedded in the slab. Such embedded fossils can be spotted and identified by a slight depression on the bedding plane, as can be seen here. The fossil can then be worked out by preparators using sharp instruments such as an X-Acto knife. The depression on this slab was formed from a specimen of the crustacean *Mecochirus,* which is embedded in the slab. It could be exposed by carefully removing three to five mm of rock to get to the fossil; such preparation is time consuming. (Value range F, unprepared).

Eryon arctiformis. An early lobster, one of the more distinctive arthropod fossils from Solnhofen. This specimen has been outlined with iron oxide (ocher) to make it more obvious. Solnhofen fossils sometimes are not too obvious and are thus highlighted. Black splotches to the right are from manganese dioxide crystals (pyrolusite), a common mineral seen on many Solnhofen fossils. (Value range E).

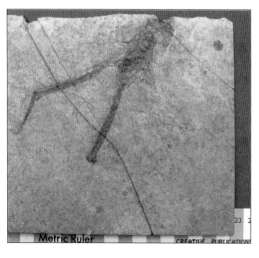

Mecochirus longimanus. These "long armed crabs" are one of the more frequently found fossils at Solnhofen. This specimen, although not too well preserved, has original carapace material on it. (Value range F).

Denderites

(Manganese dioxide crystals forming a fractal pattern) associated with the Solnhofen Plattenkalk

Denderites appear particularly abundant in association with fossil fish of which many occur in the Solnhofen Plattenkalk.

Gyrodus hexagonus. Part and counterpart of a deep-bodied fish fossil from Solnhofen. The fish impression is covered with dendrites, a fractural pattern made by crystals of manganese dioxide (pyrolusite), which grew in the vicinity of the fossil. Similar dendrites can be seen along a crack (joint) in the slab where the dendrites radiate out from the crack. (Value range E).

Gyrodus sp. Another deep-bodied fish from Solnhofen with numerous manganese dioxide crystals radiating out from it. Washington University, Dept. of Earth and Planetary Sciences Collection.

Close-up of dendrites along the crack in the previous photo. Solnhofen Plattenkalk, Solnhofen, Germany.

Fossil arthropods are sometimes found embedded in the interior of a slab of plattenkalk and can be beautifully preserved under these conditions but they have to be carefully worked out.

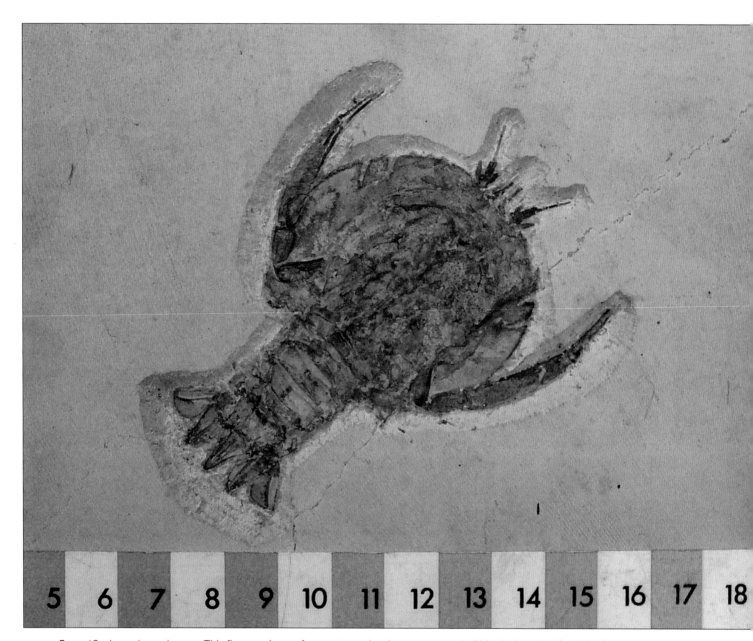

Eryon (Cycleryon) propinguus. This fine specimen of a crustacean has been preserved within the interior of a slab of plattenkalk. A 2-3 mm layer of overlying limestone has been scraped away in the preparation of this specimen. Such specimens are spotted by their characteristically shaped depressions by either quarry workers or stone masons who then set them aside for preparation and/or sale. (Value range D).

Arthropods-Insects

Solnhofen Plattenkalk is known for its fossil insects, although these are usually not as well preserved as are the insects, other paleontologic windows such as the Lower Cretaceous Crado Formation of Brazil.

Cymatophleba sp. This dragonfly is the most frequently found Solnhofen insect. Such insect fossils are not common in the plattenkalk, which is a marine deposit. Those insects found represent those which ventured out over the lagoons and didn't make it back to land. The dark bands on the slab are manganese oxide darkened cracks or joints. (Value range D).

Cymatophlebia sp. Another dragonfly, this example has been slightly highlighted with clear shellac. (Value range D).

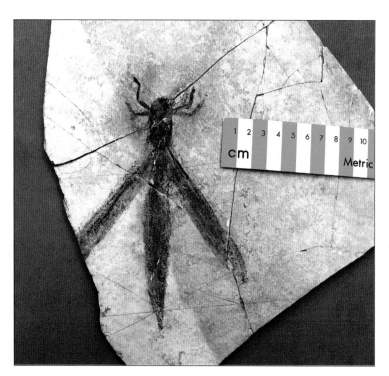

This is one of the largest insects found at Solnhofen. It is a dragonfly preserved in an odd position. (Value range D).

One of the many outcrops of the slabby, Weissen Jura limestone exposed in the quarries near Solnhofen.

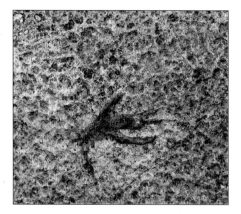

"Cricket." An early occurrence of this insect. Usually Solnhofen insects are not well preserved on the bedding planes of the Plattenkalk and are also rare. They are often preserved in the manner shown. (Value range D).

Solnhofen Plattenkalk quarries, 1993: Most of the stone quarried here is used in interior and exterior construction. Solenhofen Plattenkalk is also used in the process of lithography, its fine, homogeneous texture being particularly well suited for this stone-based printing process.

The author (middle) at one of the Solnhofen quarries in 1993.

Vertical wall of Plattenkalk. The smooth rock face of this quarry wall is a consequence of the presence of a large joint (or crack) in the rock. Such a joint forms a smooth quarry face like this when rock being quarried encounters the joint. Such joints make logical places to end quarrying operations, producing a smooth, stable face.

These are insects from another Jurassic (and Lower Cretaceous) paleontologic window, that of Liaoning Province, China.

Ephemeropsis trisetalis, a may fly. This is the most abundant megafossil in the Jurassic lake deposits of Liaoning Province, China. The Liaoning deposits are about the same age as the limestone of Solnhofen, but are fresh water rather than marine, as is the limestone at Solnhofen. It is often difficult to determine the Jurassic-Cretaceous boundary in sequences of sediments like that which produces the fossils of Liaoning Province. Some references have most of the fossils from this area as Lower Cretaceous in age; others cite a Late Jurassic age for most of them. One problem is that this thick sequence of fossil bearing strata spans the Jurassic-Cretaceous boundary. (Value range F).

Ephemeropsis trisetalis. A group of mayflys. Liaoning Province, China. (Value range F, single specimen).

2   3   4   5   6   7   8   9   10   11   12   13   14   15

Vertebrates-Ray-finned Fish

Some of these are ray-finned (paleoniscid) fish from thin-bedded, Solnhofen-like limestone of Montana. *Courtesy of John Stade.*

Hulettia americanus. A paleoniscid fish from Jurassic rocks of Montana. Paleoniscids are the ancestors to teleosts or bony fish which for the first time can be locally abundant fossils in the Jurassic. Sundance Formation, Carbon Co., Montana. (Value range E).

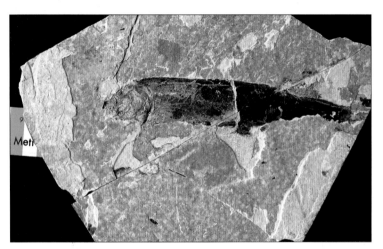

A paleoniscid fish from Late Jurassic (sometimes given as earliest Cretaceous) lake deposits of Liaoning Province, Manchuria, China. (Chaomidianzi Formation). (Value range E).

Hulettia americanus. A paleoniscid fish preserved in slabby limestone somewhat reminiscent of the Solnhofen Plattenkalk. Slabby, fine grained limestone is characteristic of Jurassic marine rocks in a number of areas of the world, including the western U.S. Sundance Formation, Carbon County, Montana.

Dapedius punctatus. A holostean (ganoid) fish from Holzmaden, Wuerttemberg, Germany. Holzmaden is another Jurassic paleontologic window. *Courtesy of Dept. of Earths and Planetary Sciences, Washington University, St. Louis.*

Vertebrates-Bony Fish (teleosts)

Teleosts (bony fish) become somewhat common in the Late Jurassic and the Cretaceous and are the dominant fish today in both marine and fresh water; they descended from the ray-finned fish, also locally common in Jurassic rocks.

Leptolepis (leptolepides) sprattiformis. A distinct, well preserved specimen of this early teleost. Solnhofen Plattenkalk. (Value range E).

Fossil Fish
Lycoptera species
Upper Jurassic
Lycoptera Layers
Liaoning Prov, China

Lycoptera sp. A teleost or bony fish from lake deposits of Liaoning, northeastern China. These fish are one of the most common fossils from this prolific lagerstatten or paleontologic window. (Value range G).

Solenhofen, Bavaria

Leptolepis (Leptolepides) sprattiformis. These primitive bony fish (teleosts) are one of the most common fossils of the Solnhofen Plattenkalk. (Value range F, single specimen).

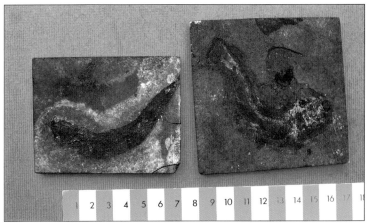

Leptolepis sp. An early teleost fish. This is a small bony fish similar to those found at Solnhofen. They are preserved in a thin, slabby organic rich shale. Portugal.

"Kugelfish." A deep bodied bony fish from a thin limestone bed of the famous middle Jurassic shales of Holzmaden, Baden-Wuerttemburg, Germany. (Value range F).

Vertebrates-Turtles

This spectacular fossil turtle is from the Late Jurassic paleontologic window of Liaoning, northeastern China.

Manachurochelys manchouensis Endo and Shikama 1942. A superb fresh water turtle with preserved soft parts. Soft tissues such as the tail, head, and legs of this soft-shelled turtle have left their presence on this fossil as a film of iron and manganese oxides. These oxides are believed to have been concentrated by bacterial action during the process of decomposition of the soft parts of the turtle. In this process the decomposers "fed" on the proteins of the animal after its burial in sediment, a process that attracted iron and manganese oxides from the surrounding sediments. Such soft-bodied preservation is rare but can be found associated with lake deposits such as that of Jurassic (and early Cretaceous) strata in Liaoning Province, China, and insects from the Crato Formation of Brazil. Specimen from Liaoning, Manchuria, Northeastern China. (Value range B).

This is a freshwater turtle similar to those living today.

A fresh-water turtle (*Trionyx*). Liaoning Province, northeastern China. This is a widely occurring fresh water turtle, the genus is still living today.

Vertebrates,
Ichyosaurs and Plesiosaurs

These are the best known marine ruling reptiles of the Jurassic.

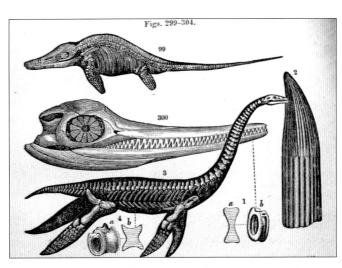

Ichthyosaur (top and middle), Plesiosaur (bottom), from Charles Lyell, *Elements of Geology*, 3rd edition.

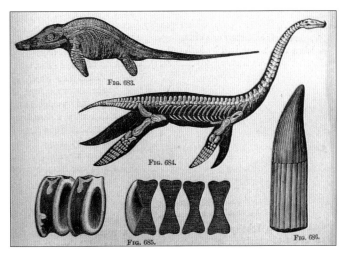

Similar illustrations as in previous photo of an ichthyosaur and a plesiosaur, also from Lyell, *Elements of Geology*, 2nd edition, John Murray, Publisher.

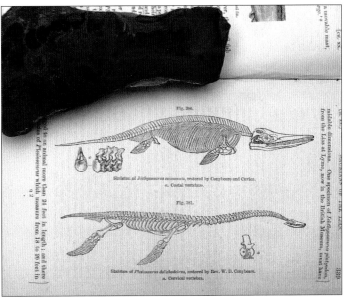

Icthyosaur fragment from Lyme Regis, Dorset, England, holding down an illustration of an ichthyosaur, which also came from Lyme Regis. From C. Lyell's *Elements of Geology*.

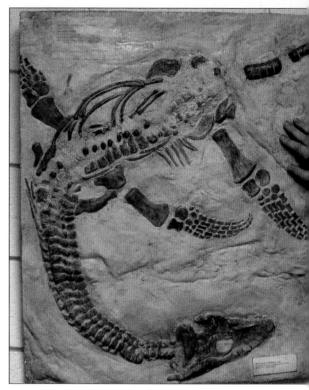

(*Plesiosaurus*) cast. This is the other large saurian originally found in rubble from a collapsed sea cliff at Lyme Regis, England, in 1823. Plesiosaurs were documented later by larger specimens of a type having very long necks, sometimes described as an animal looking like a snake threaded through the body of a turtle. This short necked plesiosaur was found complete in a large slab of grey shale and like the Ichthyosaur of the previous photo served to reinforce the concept of extinction of higher life forms over the megatime of the geologic past. These fundamental concepts, so basic to an understanding of the history of life, were unfolding in the early nineteenth century. Understanding of this basic information would lead to the concept of natural selection and the pivotal part it played in evolution. The original specimen from which this cast was made, and is now in the British Museum, was collected by Mary Anning. Casts of this specimen were made and distributed by David Jones of Jones Fossil Farm, Worthington, Minnesota. (Value range E).

An 1830s reconstruction showing icthyosaurs, plesiosaurs, and pterdactyals as interpreted by Henry Thomas De la Beche in one of the earliest representations of a Jurassic seascape in what would become Lyme Regis, England. This mural is now on display within the British Museum.

Plastic model of a plesiosaur, one of the "dinosaurs" that often come with dinosaur models and, of course, is **not** a dinosaur.

Ichthyosaurus communis. A fierce winter storm in 1823 caused the collapse of sea cliffs at Lyme Regis in southern England, which, even at the time, was well known for its fossils. A young lady by the name of Mary Anning made her living by collecting and selling seashells and fossils to seaside tourists. Among the fallen slabs she found the fossil of a large, lizard-like animal that was purchased from her and ended up in the British Museum. This was one of the fossils that gave strong support to the concept of life forms, now extinct, once having existed in the geologic past. Fossils found in rock strata like that at Lyme supported the idea of life having changed over immense spans of time. Such ideas, in conflict with strict religious interpretation at the time, were being suggested by the study of rock strata and connection of such rock strata to immense spans of time. This is a cast of the original specimen found by Mary Anning. Specimen through David Jones of Jones Fossil Farm, Worthington, Minnesota.

Plesiosaurus sp. A life-sized model of a Plesiosaur displayed at the 2008 MAPS EXPO fossil show in Macomb, Illinois.

Reptiles-Pterosaurs

Probably no group of fossils lends itself less to amateur, participatory paleontology than do pterosaurs. Fossils of these flying reptiles are truly rare and when they do occur their bones can be very hard to recognize. Most pterosaurs are known only from individual bones, articulated specimens being quite rare. Nevertheless, the prospect of finding pterosaur fossils can be a strong incentive to collecting that some persons find irresistible, like hoping to win the lottery.

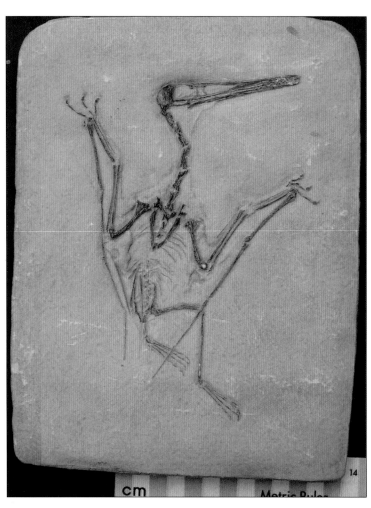

Pterodactylus sp. Complete pterodactyls are one of the most prized fossils of the Solnhofen Plattenkalk of Germany. This replica (cast) is from a particularly nice specimen from Solnhofen and has been widely distributed, as casts. (Value range F).

The **cast** of another species of pterodactyl from the Solnhofen Plattenkalk. (Value range E).

Rhamphorhynchas sp. A somewhat disarticulated pterodactyl specimen from the Solnhofen Plattenkalk. *Courtesy of Washington University, Dept. of Earth and planetary sciences collection.*

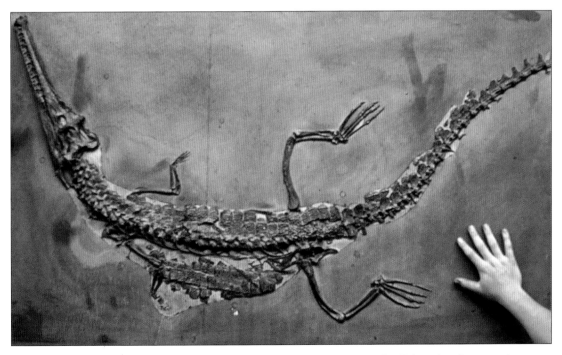

Steneosaurus sp. Teleosaur or sea crocodile. Lower Jurassic, Posidonienschiefer, Holzmaden, Germany. A marine, lizard-like reptile (saurian) from the celebrated Holzmaden shales. Washington University, St. Louis collection. This specimen comes from one of the Jurassic localities of southern Germany, the Lower Jurassic Posidonienschifer of Holzmaden. A hard, black organic-rich shale ("slate") quarried near the town of Holzmaden has yielded complete saurians like this and is probably best noted for its complete ichyosaurs. Other than ammonites and belemnites, which are common in the shales, most Holzmaden fossils are quite pricey and are seen in many museums, at least in those well endowed museums that can afford one.

Vertebrates-Tracks

These are tracks of dinosaurs. All of them came from strata of the northeastern states and were originally considered as Triassic in age.

Anomoepus intermedius Hitchcock (cast). Newark Group, Turners Falls, Massachusetts. This genus of dinosaur trackway occurs in strata exposed at Turners Falls and such strata has been found to be Lower Jurassic in age rather than Triassic as originally described by Edwin Hitchcock in the mid-nineteenth century.

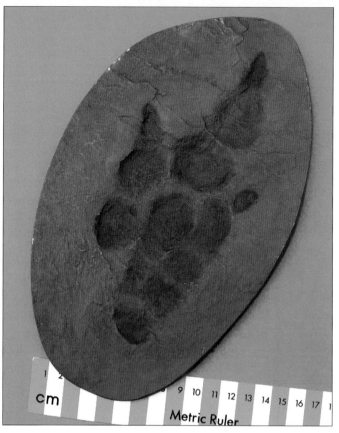

Grallator sp. (cast). This ichnospecies of dinosaur track has been determined to be early Jurassic in age. The Newark Group, in which it occurs, has been found, from associated fish fossils, to be of early Jurassic age rather than Triassic. Turners Falls. Massachusetts. (Value range F).

Anomoepus curvatus Hitchcock (left rear footprint). This ichnospecies is similar to that of the previous photo. There has been a tendency for paleontologists to give different species names to variants of a fossil, which may or may not be biologically valid. This is because there is no way to really know if an organism, represented by a fossil such as a shell or a skeleton, is really a valid biological species or is just a individual variation within a population. This really becomes a problem with fossil tracks as these can vary with the consistency, type, and walking stance of the organism making the track.

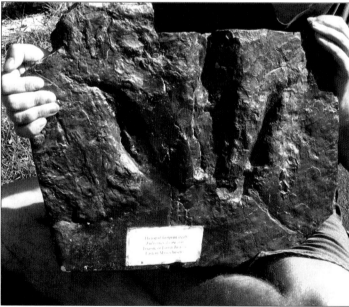

Eubrontes divaricatus Hitchcock. Newark Formation, Upper Portland Series, Chicopee Falls, Massachusetts. The Newark Series (or formation) has been found from microfossils, fish, and plants to be early Jurassic in age rather than Triassic as it was originally designated. This is also a large dinosaur track, especially for the Triassic as Triassic dinosaurs were relatively small. Large dinosaurs, such as the example that made this track, are characteristic of the Jurassic and Cretaceous, but are **not characteristic of** the Triassic. Dinosaur tracks in the Connecticut River Valley this large, however, are relatively rare. (Value range C).

Vertebrates-Dinosaurs

Particularly large dinosaurs lived in the Late Jurassic and their bones, usually found in a fragmented condition as surface finds, can be locally common in Jurassic strata of the western states, particularly in the Late Jurassic Morrison Formation.

This is a slice through a large dinosaur bone from the Upper Jurassic Morrison Formation. Dinosaur bone can be relatively abundant in the Morrison, particularly fragments of large bones. The Morrison Formation of Wyoming, Colorado, New Mexico, and Utah contains some of the largest concentrations of dinosaur bone in the world. The Morrison Formation is also a source of uranium minerals such as carnotite and autunite and these minerals may be associated with dinosaur bones. Dinosaur bone concentrations, along with large carbonized logs, have been one of sources for uranium in the U.S. over the past five decades. Uranium minerals concentrate in the bone as uranium is attracted to calcium phosphate, the mineral component of the bone. In the process of mining uranium, many nice (some possibly complete) dinosaur skeletons have been sent to the rock crusher. Their radioactivity, however, in mounted dinosaur skeletons, might make them somewhat scary as museum reconstructions. (Value range F).

Another large chunk of dinosaur bone from the Morrison Formation. Such large chunks have been widely distributed through rock hound activity and when the bones are agatized they are used in making dinosaur bone jewelry. Polished pieces of this agatized dinosaur bone are sold by fossil and natural-history dealers as "gembone."

"Gembone." Agatized fragments of dinosaur bone from the Morrison Formation are often cut into cabochons and made into dinosaur bone jewelry. Rock hounds have been collecting chunks of this bone for decades. Many (or most) of these bone fragments are believed to have come from large sauropods like *Apatosaurus (Brontosaurus).* (Value range G, single specimen).

Excavations of the bone bearing rock face of Dinosaur National Monument in 1989; excavation is an ongoing process at "Dinosaur." Part of this outcrop is the same face as in the previous photo, but with additional exposed bones.

One of the largest concentrations of well preserved dinosaur bones was found in the early twentieth century near Vernal, Utah. The site was worked by Earl Douglas of the American Museum of Natural History, also early in the twentieth century. Later, this site became Dinosaur National Monument where on-going excavation of a hillside outcrop (now covered by a large building) has progressed over the decades. This is a photo of part of the bone-bearing outcrop as it appeared in 1960.

Mounted "sauropod" skeleton (*Apatosaurus*) made around 1915 from bones excavated in the Earl Douglass Quarry and now in the Chicago Naturàl Museum (Field Museum). *Courtesy of Chicago Museum of Natural History.*

Dinosaur gastroliths or "stomach stones."

Colorful plastic models of "Brontosaurus" as well as other dinosaurs have been popular kids toys since the 1950s.

Gastroliths. These highly polished stones found in the Morrison Formation and other strata are associated with concentrations of dinosaur bones. In a few instances, concentrations of them have been found directly associated with a dinosaur rib cage. Such stones are considered to be gastroliths, that is stones swallowed by dinosaurs (and perhaps crocodiles as well) to assist in the grinding of woody vegetation in the animals stomach. This was a tactic similar to that used by birds of today, which grind up food within their stomachs with swallowed stones. That is the reason why chickens eat small rocks!

A life sized *Apatosaurus (Brontosaurus)* at Dinosaur Park, Rapid City, South Dakota. Life-sized dinosaur models represent a popular paleontological genre, particularly in the western U.S. and Canada.

Dinosaur Models

As dinosaur bones and other large fossils of ruling reptiles are one of the least logical fossils to collect, dinosaur models are a more suitable item; and pictures of some large models may be even more logical as life-sized models like dinosaur bones themselves can also be a bit inconvenient as a collectable!

Apatosaurus (Brontosaurus) appeared not only as the Sinclair Oil company's logo starting in the 1930s, but surfaces serendipitously as the ubiquitous American (and Canadian) "Advertising Sauropod." These images were particularly popular from the 1950s to the '80s.

Life size *Brontosaurus (Apatosaurus)* model made of metal and exhibited outdoors at a state prehistoric life and fossil museum at Vernal, Utah.

Dinosaur Trackways

The least logical (and usually illegal) fossils to collect would probably be large dinosaur tracks and track ways like these. A camera and the right angle of sunlight is the proper method to "collect" such fossils. The angle of sunlight, usually a low one, can make a big difference in seeing such trace fossils as well as in photographing them.

Apatosaurus (Brontosaurus) Trackway. Trackways of large, quadruped dinosaurs occur on bedding planes of sandstone of the Morrison Formation along Colorado's Purgatory River in Picketwire Canyon. A particularly spectacular but staged view of these same trackways are in the January 1993 article on dinosaurs in *National Geographic Magazine. Photo courtesy of Matt Forir.*

Trackways which may have been made by *Allosaurus,* a carnosaur associated with *Apatosaurus* and found in the Late Jurassic Morrison Formation. *Allosaurus* may have preyed upon *Apatosaurus* as the bones of both of them are found together in the Morrison Formation. These trackway photos, examples of fossils collected by photography, are courtesy of Matt Forir.

Another view of what are believed to be brontosaur trackways in Picketwire Canyon, Colorado. The upper part of the hill in the background is composed of the Dakota Sandstone, a rock sequence which forms the base of the Cretaceous System in much of western North America. The trackways were made in sand (now sandstone) deposited by braided rivers during the latest part of the Jurassic Period. *Photo courtesy of Matt Forir.*

Grinning Stegosaur! This rather peculiar, life sized "dino" is an example of dinosaur "folk art," which appears sporadically in the western U.S. and Canada. They are usually made of fiberglass, a material that holds up well under the sun in a harsh climate.

Dinosaur coprolites?

Another view of the same trackway as shown previously.

More fiberglass dino's—the previous ones just weren't enough!

"Kids toys." Stegosaurus models.

Geode-like masses found in the Morrison Formation are considered by some to be mineralized coprolites, possibly coprolites derived from the droppings of large sauropods such as *Apatosaurus (Brontosaurus).* This attractive, polished sphere has been made from such a geode-like mass. Spheres and polished masses of this material show up at rock shows with some frequency. The making of spheres is a labor intensive process so they can be rather pricey. (Value range E).

Geode-like quartz masses which some paleontologists have considered to be mineralized dinosaur droppings (coprolites). These are found associated with stream deposits of the Upper Jurassic Morrison Formation and are often associated with dinosaur bones and gastroliths. On the outcrop they can resemble a large cow patty; however, whether they are actually coprolites is debated. This polished specimen is from the Morrison Formation of southwestern Oklahoma. (Value range F).

Birds or Feathered Dinosaurs?

Actual fossils (not casts) of these are even rarer
than are those of pterosaurs.

Archeopteryx lithographica. On the left is a photograph of the original Berlin specimen found in 1865 at Solnhofen, Bavaria. This fossil served to reinforce Charles Darwin's natural selection based theory of evolution, as it is an intermediary between reptiles and birds; such intermediaries are quite rare in the fossil record. On the right is a cast of a replica made of this famous fossil. These replicas have been distributed by the Internet.

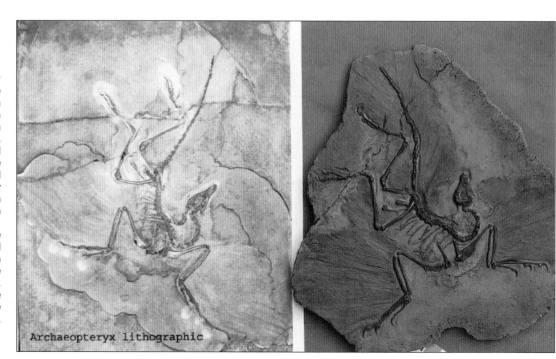

A closer look at one of the carved replicas of *Archeopteryx lithographica*. A "cast," actually a cast of a carving of what some consider to be the world's most valuable fossil, a superb specimen of a toothed bird or feathered dinosaur from the Solnhofen Plattenkalk. Replicas like these, in a way, are in the same category as are replicas of the Mona Lisa or phony Rolex watches. This specimen was made in China and sold (as a cast) on the Internet. Specimens of feathered dinosaurs or toothed birds are known (in quantities greater than those from Solnhofen) from the Late Jurassic-Early Cretaceous strata of Liaoning, northeastern (Manchuria) China, but are still quite rare. *Courtesy of Steve Riggs Jones.* (Value range F).

Outcrop of massive Navajo Sandstone in central Wyoming. The Navajo and Entrada sandstones represent a series of fossil sand dunes of early Jurassic age. These beds lack fossils other than some dinosaur trackways, which occur on infrequent bedding planes. In the foreground are Triassic sandstones and siltstones of the Chugwater Formation which, on weathering, produced the sediments which makes up the red, sandy road.

Outcrop of Lower Jurassic Navajo Sandstone in the San Rafael Swell of central Utah. Red beds of the uppermost part of the Triassic Period can be seen at the very bottom of the canyon.

Outcrop of the late Jurassic Morrison Formation in central Wyoming. The Morrison Formation can locally contain a variety of fossils, which include dinosaur bones, petrified logs, and sometimes compression fossils of plants. It is often variegated or multicolored with grey, grey-blue, purple, reddish-yellow, and reds represented.

Another typical and pastel hued outcrop of the Late Jurassic Morrison Formation, Central Wyoming.

Bibliography

Case, Garard R., 1982. *A Pictorial Guide to Fossils*. Van Nostrand Reinhold Company.

Gould, Stephen J., 1980. "The Telltale Wishbone" **in** *The Panda's Thumb: More Reflections in Natural History*. W.W. Norton and Co., New York-London.

_____, 1991. "Bully for Brontosaurus" **in** *Bully for Brontosaurus: Reflections in Natural History*. W. W. Norton Co., New York-London.

_____, 1993. "Bent Out of Shape" **in** *Eight Little Piggies: Reflections in Natural History.* W. W. Norton and Co., New York-London.

Norell, Mark 2005. *Unearthing the Dragon, The Great Feathered Dinosaur Discovery.* Pi Press, New York, New York.

Norell, Mark A., Eugene S. Gaffney and Lowell Dingus. 1995. *Discovering Dinosaurs in the American Museum of Natural History*. Alfred A. Knopf.

Psihoyos, Louie and John Knoebber, 1994. *Hunting Dinosaurs.* Random House, New York, New York.

Unwin, David M., 2006. *Pterosaurs.* Pi Press, New York, New York.

Weishampel, David B., 1990. Peter Dodson and Halszka Osmolska Ed. *The Dinosauria.* University of California Press.

Chapter Five

Mesozoic "Prehistoric Animals" and World's Fairs

Modern geology, with its emphasis on fossils and megatime, had its inception in the 1820s and '30s. With this "new" geology, emphasis was on information which rocks and strata could reveal that was related to the formation and development of the planet; earlier concern with rocks and the earth focused on minerals and mining. Pioneers in this new geology start with James Hutton at the end of the eighteenth century; Hutton began investigations that confirmed the significance of sequences of sedimentary rocks (strata) in earth history. Such strata included fossils and the part they played in determining the progression of life; life revealed to have changed and often seemed to have been terminated over vast amounts of time. Besides Hutton, this endeavor included early nineteenth century geologic pioneers like William Smith, Adam Sedgwick, Roderick Murchison, and Charles Lyell. In the 1840s, came the discovery of dinosaurs and the fact that large, extinct reptiles had once existed during that part of earth history becoming known as the Mesozoic (Middle) Era of geologic time. Such large and strange life forms tweaked the public's interest in fossils and prehistoric life. The public then (as now) became fascinated by the idea of large, prehistoric beasts having once lived and inhabited the realm of deep time, being particularly attracted to those animals dubbed at the time as "saurians." This fascination with large prehistoric creatures was accompanied by significant new developments in technology, which included the railroad, the telegraph, and photography, all of which – like the prehistoric monsters being uncovered – seemed to the public of the nineteenth century to be almost magical. The 1840s in England ended with a goal of staging a major public event, an event which would highlight this new technology and show it off to the world. The display would take place in London and was to be done in such a manner that the world would be mesmerized by its brilliance and magnitude; thus was born the concept of the **World's Fair**. The envisioned event would be known as the Crystal Palace Exposition and would consist of working displays of this new technology as well as other accomplishments of the "civilized" world. The exposition would be housed in a huge glass building to be known as the "**Crystal Palace.**"

As the prehistoric creatures being discovered by the new science of geology and its offspring, paleontology, fascinated the public, it was decided that part of the exposition would focus on them, particularly the large "monsters" of the Mesozoic Era. Life-sized models of large prehistoric animals like Iguanodon and the Megalosaurus would be constructed and displayed in a variety of settings, one of which would include outdoor settings on the grounds near the Crystal Palace itself. Thus was established a precedent for presentation of prehistoric "monsters" to the public, a feature which would become a part of many world's fairs of the future.

An important concept of prehistoric life was incorporated in the Crystal Palace dinosaurs. Sir Richard Owen, the originator of the concept of dinosaurs, was a strict creationist! Owen believed that all major life forms such as reptiles, mammals as well as distinctive members of the invertebrates (phyla) were independent creations of God. With Owen, dinosaurs represented the acme of perfection of reptiles—cold-blooded animals designed by a benevolent supreme being intended for an earlier, warmer Earth. Warm-blooded mammals, on the other hand, were designed and created by God later for the cooler earth of the Cenozoic Era (the age of mammals) and today. Modern reptiles, like snakes, lizards, and turtles, Owen believed were degenerate forms of this previous peak of reptilian creation—Owen's dinosaurs. This emphasis on creation rather than transmutation (viz. some form of evolution) remained as a focus of World's Fairs influenced by the nineteenth century Crystal Palace model. The 1904 and 1906 expositions in St. Louis and Portland, in the States, had a creation exhibit as one of their major public attractions.

Megaladon from the Crystal Palace Exposition, 1851, at Crystal Palace Park, London. This was one of the first reconstructions of a dinosaur, modeled after the *Iguanadon,* the first scientifically described dinosaur. These outdoor models have survived London's climate nicely since they were made some 160 years ago. *Photo courtesy of Matt Forir.*

A group of saurians. Mosasaur (center) and long necked plesiosaur (far left), an Ichyosaur is in the background. Portland stone (Portland Cement) models made for the Crystal Palace Exposition, the first world's fair, in London, 1851.

Another view of the mosasaur (center) and the ichyosaur at Crystal Palace Park, London. *Photo courtesy of Matt Forir.*

Long necked plesiosaur (right) at Crystal Palace Park, London. Plesiosaurs were one of the first Mesozoic "ruling reptiles" to be described after the mosasaur.

Teleosaurs. These "skinny crocodiles" were first described in Germany and models of them made for the Crystal Palace Exposition, Crystal Palace Park, London. *Photo courtesy of Matt Forir.*

The second "World's Fair," one intended to equal or exceed the efforts of the 1851 Crystal Palace Exposition, was the "Centennial Exposition" of the United States in 1876. This exposition would celebrate and commemorate, in a grand manner, the centennial of the birth of the Republic. One of its goals was to outdo the technological displays of the earlier Crystal Palace Exposition, but a secondary goal was to outdo the "prehistoric monster" exhibits of the London Exposition. The same Mr. B. Waterhouse Hawkins who had prepared prehistoric animal exhibits for the London Exposition was to oversee the building of new models for a prehistoric animal exhibit in New York City's Central Park; similar models were then to be installed at the Centennial Exposition in Philadelphia.

The models would focus on **saurians**, particularly large and strange ones that were turning up in the American West as a by-product of western expansion following the Civil War. Some of these had been described for science by Joseph Leidy of Philadelphia before the war, however post war discoveries, aided by westward expansion and the railroads, seemed to eclipse earlier discoveries. Many of these post war discoveries were made by the antagonists Othniel Marsh and Edwin Drinker Cope, whose discoveries in the American West were equaled only by the amount of personal animosity they held toward each other. Models of the recently discovered dinosaurs *Brontosaurus* and *Stegosaurus* from the Jurassic Morrison formation of Wyoming and Colorado were commissioned to be made in New York City by the venerable Mr. Hawkins. They were later to be shipped to Philadelphia for display at the exposition. These were to be life-sized models that included *Triceratops* as well as large duckbilled dinosaurs found in Cretaceous rocks of Wyoming and Montana. Building of the models commenced in New York City during the scandal ridden Grant administration, an arrangement that unfortunately included the notorious "Boss Tweed" of New York overseeing financial matters. To make a long story short, Tweed managed to commandeer the allotted funds for completion of the models as well as to "get his hands" on funds intended to cover shipping costs to transport the huge models to Philadelphia. When these necessary funds disappeared, the uncompleted models were unceremoniously broken up and disposed of in New York's East River and thus ended plans to eclipse the Crystal Palace's prehistoric-model exhibits. Some dinosaur reconstructions as well as mounted skeletons of a Wooly Mammoth and North American Mastodon did become part of the exhibits at the Centennial Exposition, but this was not the great show of prehistoric monsters as originally planned.

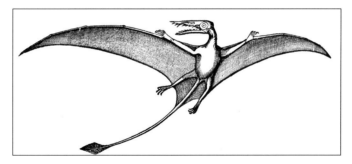

Besides dinosaurs, the Centennial Exposition in Philadelphia was to have life-sized Pterosaurs, specifically a model of one of the large pterosaurs found in the 1870s in Cretaceous chalk of Kansas.

A life-sized model of *Stegosaurus* and *Triceratops* were to be exhibited but they ended up, as did many others, being destroyed and dumped into New York's East River as a consequence of fiscal irregularities of New York City's Tweed Administration.

The next World's Fair, the 1889 Paris Exposition, saw not only the opening of the Eiffel Tower as a great technological wonder but, on a more modest scale, models of large fossil mammals found in the vicinity of Paris (found associated with gypsum beds, the source of that city's famous plaster). The Columbian Exposition of Chicago followed in 1893, which had its share of saurian models as well as impressive mounted dinosaur skeletons of new finds in the American West. Some of the discoveries of those rivals Marsh and Cope were again well represented in Chicago's Columbian Exposition. Another phenomena, however, that of ethnology and its focus on people from around the world and their "primitive" customs, overshadowed for the first time displays of prehistoric animals. The assemblage of fossils, mounted skeletons, and models of prehis-

toric animals so intrigued department store mogul Marshall Field that, after the fair, they would become the basis for a natural history museum endowed by Field. Thus was born from the 1893 Columbian Exposition the Chicago Natural History or "Field Museum" of today. The focus on ethnology, which had been minimal in previous expositions, included tribes of the "uncivilized" world, as well as focusing on the culture and traditions of the American Indian. Realization that cultures of peoples of many parts of the world were being modified and eradicated by the spread of technology—that very technology being highlighted by the exposition—resulted in cultural exhibits increasing in importance. This included the culture of the American Indian, whose culture at the time was at last, becoming more highly regarded and appreciated by American society at large.

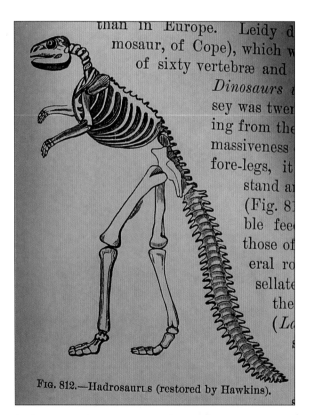

FIG. 812.—Hadrosaurus (restored by Hawkins).

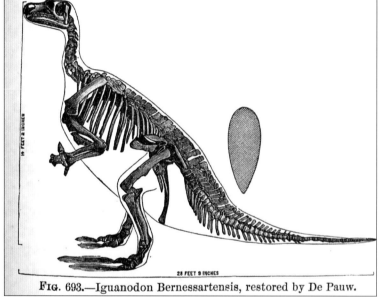

FIG. 693.—Iguanodon Bernessartensis, restored by De Pauw.

The Iguanodon. Extensive numbers of complete dinosaur bones had been found in a Belgian coalmine, which formed the basis of these dinosaurs modeled and displayed in the 1889 Paris Exposition. The nearly complete skeletons found enabled the construction of complete mounts of the bones of this earlier-discovered dinosaur. These mounts are still part of dinosaur displays in the Belgian Natural History Museum in Brussels.

B. Waterhouse Hawkins had restored dinosaur skeletons found in England in the late 1860s. He was to oversee a display of dinosaurs in New York City to be placed in Central Park and then some of the models were to be sent and displayed in the 1876 Centennial Exposition in Philadelphia.

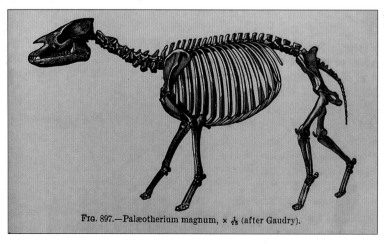

FIG. 897.—Palæotherium magnum, × ⅒ (after Gaudry).

Life-sized models of large, prehistoric mammals like these were part of prehistoric animal displays at both the 1889 and 1900 Paris Expositions. The Palaeotherium was one of the large, extinct mammals associated with Parisian gypsum deposits.

The Paris Exposition of 1900 repeated previous French exhibits with its mounts and models of some of the early mammals associated with its famous plaster, as well as high-lighting finds of prehistoric man, which included Cro-Magnon and Neanderthal man—two "missing links" recently found in Europe. Next in line and not to be outdone by Chicago's Columbian Exposition was an exposition to take place in the city of St. Louis, Missouri. St. Louis was able to compete with Chicago as its population in the 1880s and '90s had expanded considerably as a consequence of European immigration and the city decided to put on its own show, a show that would outdo Chicago's Columbian Exposition. It would be known as the "Louisiana Purchase Exposition" and would commemorate President Thomas Jefferson's land purchase of 100 years earlier. One year late, as was the Columbian Exposition (Columbus-1492; Louisiana Purchase-1803), the 1904 "St. Louis World's Fair" would be the **greatest yet**. The Louisiana Purchase Exposition with its ethnological exhibits scooped anything that had been done previously; entire tribes were transported from their homeland, which included entire villages of "Eskimos" and a complete tribe of Igorrote "Indians" from the Philippines. Prehistoric animal displays were eclipsed by the ethnological ones but did include a mounted *Triceratops,* a *Stegoceras* skeleton, a prehistoric whale as well as mounted skeletons of ice age animals (Pleistocene megafauna) like the mammoth and mastodon. Related also to prehistoric animals and geology were trips to sites in Missouri by train which would allow fair goers to experience a prehistoric animal dig first hand, such as the dig at the Kimmswick mastodon site south of St. Louis (The Koch "Fossil Farm" and now Missouri's Mastodon Park) as well as visits to some of the caves of the Ozarks.

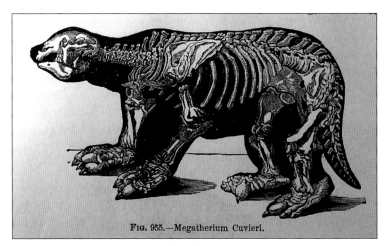

FIG. 955.—Megatherium Cuvieri.

The *Megatherium* was another huge, extinct mammal displayed at the 1889 Paris Exposition. Its species name, *cuveri,* was after Georges Cuvier, who was the first to accurately identify and reconstruct extinct vertebrates from fossil bones. Of the work he did, at the time it was proclaimed to be "putting the dead back to life," a feat that took place before the "discovery" of dinosaurs by Richard Owen of England.

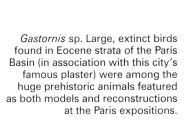

Gastornis sp. Large, extinct birds found in Eocene strata of the Paris Basin (in association with this city's famous plaster) were among the huge prehistoric animals featured as both models and reconstructions at the Paris expositions.

Exhibits relating to the antiquity of man formed part of both the 1889 and 1900 Paris expositions. Excavation of bone caves like the one illustrated here yielded bones of fossil man, which included Cro-Magnon and Neanderthal man.

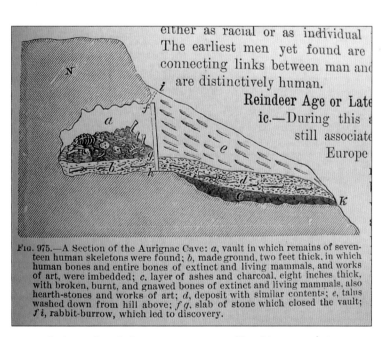

FIG. 975.—A Section of the Aurignac Cave: *a*, vault in which remains of seventeen human skeletons were found; *b*, made ground, two feet thick, in which human bones and entire bones of extinct and living mammals, and works of art, were imbedded; *c*, layer of ashes and charcoal, eight inches thick, with broken, burnt, and gnawed bones of extinct and living mammals, also hearth-stones and works of art; *d*, deposit with similar contents; *e*, talus washed down from hill above; *f g*, slab of stone which closed the vault; *f i*, rabbit-burrow, which led to discovery.

Aurignac Cave in France was a source of fossil human remains, which formed the basis of an exhibit at the 1900 Paris Exposition. Such bone caves, with their ancient human remains, paralleled the emphasis on ethnology, which was emphasized during world's fairs of the early twentieth century.

Mentone Cave fossil? Human skeleton. Discovery of human skulls that were anatomically different from those of modern man in French caves (and elsewhere) were subjects of the Paris Expositions. Mentone Cave, near Nice, France, yielded this skeleton, which was exhibited at the 1900 Paris Exposition along with similar human skulls of Cro-Magnon man, now considered as a sub-species of modern man.

Life-sized model of a Columbian Mammoth at the 1893 Columbian Exposition in Chicago. In the background can be seen mounted dinosaur skeletons and early mammals, which now form part of the Field Museum collections.

Chicago's Field Museum has become one of the prime natural history museums in the world and is an icon of Chicago. The museum has "launched out" into other locations in Chicago as well. Here is a life-sized replica of *Brachiosaurus,* a Jurassic dinosaur which oversees one of the terminals of Chicago's O'Hare Airport.

World's fairs exuded nationalism on a massive scale, each nation both trying to outdo those that had gone before them in the presentation of new technology and also comparing the nation's technology with that of other parts of the "civilized" world. Here a very nationalistic Miss Liberty guards a plinth mounted Graphophone (phonograph) of late nineteenth century vintage. The phonograph was one of many marvels that would appear on the scene at world's fairs following those marvels showcased at the Crystal Palace Exposition – the telegraph, photography, and steam powered boats and trains.

Not to be forgotten in the realm of geology was "Creation," a major attraction on the Pike which took fairgoers through a large plaster-of-Paris building by means of boats where participants would view dioramas representing the "days" of creation of the Bible (Genesis I). In "Creation" each room represented one of the Bible's Six Days of Creation. The boat visited this spectacle—one room at a time—the sequence starting with day one and the darkness of primeval earth and ending with the creation of man (room six and day six).

The time scale represented here was the 10,000 years of young earth creationists who at the time had an ally with the English physicist Lord Kelvin, whose calculations regarding the "age of the Earth" was based upon Kelvin's field of thermodynamics. With thermodynamics Kelvin had determined from measurements of the internal heat of the Earth that the 10,000-year-old earth of the Creationists was a scientific possibility, a billions of years old earth, derived from geology, was, according to his calculations, an impossibility! Concurrent with this young earth perspective was the radium exhibit in the Palace of Liberal Arts. Here, in a small room, the walls of which were coated with the phosphor zinc sulfide, the public could experience the mysterious emanations of newly discovered radioactivity, which emanated from the radium. A small speck of this recently discovered element caused the walls of the room to glow a brilliant green as radioactivity—given off by the radium—excited the phosphor, causing it to glow like a modern TV screen. This exhibit, not too far from the Pikes "Creation," was ironic because radioactivity and radiometric age dating in the future would be the vehicle which would issue the final "death blow" to a scientifically documented young age of the Earth. Radioactivity also would be found to be the source for the thermal energy of the Earth's interior, a discovery which negated the calculations from which Kelvin had "scientifically proven" the young Earth.

Ice age mammals, such as Mammoths and Mastodons, excited the public's interest, as did extinct Mesozoic animals like dinosaurs. A "turn of the century" article on prehistoric animals (Mastodons) found south of St. Louis, Missouri, wetted the public's appetite for exhibits and digs of prehistoric animals. The "fossil farm" or Koch Mastodon site was opened and numerous bones collected, which could be seen (and sometimes purchased) by participants who visited the site from excursions by train. These trains left the fair grounds at regular intervals. This was one of many excursions of a geologic nature accompanying the Louisiana Purchase Exposition.

One of the large "Palaces" at the Louisiana Purchase Exposition. This was the Palace of Mines and Metallurgy, which housed various mining exhibits, including a model coal mine as well as numerous displays of minerals and other geologic phenomena. Its emphasis on mining did not include fossils and prehistoric life; these were displayed in a somewhat limited way in a similar, large building, the "Palace of Liberal Arts."

Newspaper articles like this appeared at the turn of the twentieth century, which "tweaked" the publics interest in prehistoric animals prior to the Louisiana Purchase Exposition. During the exposition, the public was offered excursions to the Mastodon site (fossil farm) as well as to various caves in Illinois and Missouri.

"Creation." The huge "maw" of this staff (plaster and straw) building led to a series of rooms, each room portraying one of six days of Creation poetically outlined in Genesis, the first book of the Bible. A 10,000 year old Earth was inferred in "Creation," an age that conflicted with the hundreds of millions of years old Earth of geology and paleontology. Ironically, in the nearby "Palace of Liberal Arts building" was presented the radium (and radioactivity) exhibit. The understanding of radioactivity and radiometric age dating, still a decade away, would become the "death knell" for the "science" of the Young Earth Creationists.

A black and white view of part of the extensive "Pike" of the Louisiana Purchase Exposition.

"The Pike." The most popular part of the Louisiana Purchase Exposition, the Pike had a variety of amusements such as a ride called "To the North Pole and back" as well as "Creation." It also saw the first display of Baby Incubators (with real live babies).

"Down the Pike at the St. Louis Exposition." A variety of musical and other sound recordings highlighting the fair were offered for both disk and cylinder Talking Machines. This one mentions the baby incubators (with crying babies) and Creation, both of which were located on the Pike. Note the "crowing" about winning Grand Prizes at both the Paris (1900) and St. Louis Expositions (1904).

Concern with ethnography reached its peak in the early twentieth century and became a major theme of world's fairs at the time. Edison's Phonograph in the 1890s effectively documented vanishing songs and dialects of Native Americans and a few years later this "reenactment" of Native America (Navajo) songs became a lively selling number in the record catalog of Edison's National Phonograph Company. Opening comments on the cylinder admonish listeners not to regard these songs as "The improvisations of a savage mind." Ethnographic recordings were made extensively on cylinders in the early 1890s by the federal department of ethnology, an agency assisted by dinosaur worker Othniel Marsh, who played an active role in it before his death just before the close of the nineteenth century.

St. Louis Tickle and St. Louis Rag. Rags written especially for the St. Louis exposition. Some persons have queried as to the reason for the introduction of old (antique) record labels in a work on fossils. The author confesses to having another collecting interest in the form of antique phonographic items. One of the notable aspects of collecting these two distinct (but not always incongruous) collectables is how they both relate to the "flow" of time. Recording technology, like life itself, "evolved" through time and both have left an interesting "record" of time dependent changes. Old records and phonographs exist on a time scale that we experience. One hundred year old records and talking machines represent a time scale readily comprehensible; fossils, on the other hand, collected 100 or even 1,000 years ago are the same as those collected today. That's because the ages of fossils are reckoned on the scale of geologic time and this is a time scale which we don't experience. The order of magnitude of the ages of most fossils is many times greater than the age of manmade things, even ancient manmade objects like stone tools, which, though a million or two years old, are still quite young when compared to the age of most fossils.

Ethnographic concern and the romanticizing of the American Indian became features of the late nineteenth century, features which eclipsed prehistoric animal exhibits at world's fairs at the end of the nineteenth and beginning of the twentieth centuries. Here a traditionally attired Native American enjoys a calumet smoke along with the sounds emanating from a open horn Gramophone. The rest of the statement "Music Hath Charms" was "Music Hath Charms **To Sooth the Savage Breast.**" It should be mentioned here also that this record label was printed by color lithography. This process prints from Jurassic Solnhofen Plattenkalk stone plates and was a popular form of color printing in the late nineteenth and early twentieth centuries.

A group of Native Americans in authentic dress pose for a picture at the Louisiana Purchase Exposition.

Ethnographic exhibits at the Louisiana Purchase exhibit eclipsed those of prehistoric animals. In some ways, the St. Louis world's fair was the first one to somewhat de-emphasize prehistoric creatures of the geologic past. This may, in part, have been a consequence of major "movers and shakers" involved in the fair lacking a personal interest in this subject. Without question, ethnographic topics took a preeminent position at this fair, one of the largest of any of the world's fairs. Ethnographic exhibits of the Louisiana Purchase Exposition included hundreds of Igorrote "Indians" from the Philippines who lived in their own huts and engaged in traditional pursuits, which included the eating of dogs. These Philippine natives are preparing a dog while fair participants ogled the event. The area of St. Louis where the Igorrote "Indians" were housed and engaged in this activity is now known as "Dogtown."

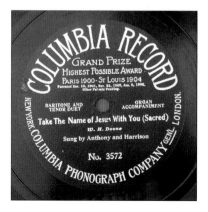

A label on a Gramophone (disk) record heralding the World's Fairs (and medals awarded to the Columbia Graphophone Co.) in the Paris (1900), Buffalo (1901), and St Louis expositions.

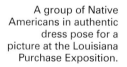

The Pan American Exposition in Portland, Oregon, had some of the same exhibits and focus as did the St. Louis Exposition. Again, besides technology, it had an emphasis on ethnology, particularly the cultural changes being forced upon of the "red man." After these expositions, focus on both prehistoric animals and threatened cultures diminished, particularly with the world's fairs of the 1910s and '20s. However, the 1939 World's Fair in New York, an event that saw the first introduction of commercial television, also saw a spectacular array of life-sized dinosaur models sponsored by the Sinclair Oil Co.

Dinosaur models, also sponsored by the Sinclair Oil Co., appeared again with the 1965 New York fair, but otherwise in many ways it was the end of an era, an era initiated in the 1850s by the "Crystal Palace Exposition" in London.

Bibliography

Burchfield, Joe D., 1975. *Lord Kelvin and the Age of the Earth.* Science History Publications, New York.

Gould, Stephen J., 1985. "False Premise, Good Science" **in** *The Flamingo's Smile, Reflections in Natural History.* W. W. Norton and Co., New York-London.

_____, 1991. "The Dinosaur Rip-off" **in** *Bully for Brontosaurus, Reflections in Natural History.* W. W. Norton and Co., New York-London.

Jaffe, Mark, 2000. *The Gilded Dinosaur: The Fossil War between E. D. Cope and O. C. Marsh and the Rise of American Science.* Crown Publishers, New York.

Rudwick, Martin J. S., 1992. *Scenes from Deep Time.* University of Chicago Press. Chicago and London.

Wilford, John Noble, 1986. *The Riddle of the Dinosaur.* Alfred A. Knopf, New York.

Apatosaurus (Brontosaurus) became famous as the logo for the Sinclair Oil Company, the organization that sponsored the dinosaur model machines of the 1965 New York World's Fair. The large dinosaur represented is now known as *Apatosaurus* and fragments of its bones can be relatively common fossils in the Jurassic Morrison Formation of the western U.S.

These dinosaur models, from dinosaur exhibits at the 1965 New York World's Fair, show that interest in prehistoric animals at world's fairs in the mid-twentieth century had not totally reached the "end of an era." These dinosaur models came from injection molding machines, which, upon insertion of a quarter, would spit out a specific dinosaur from each machine. Sponsored by the Sinclair Oil Company, the dinosaur extruding machines toured the country after the fair had closed, where they dispensed around the US numerous examples of these collectable dinosaurs.

Scene from deep time! A Jurassic seascape. Scenes depicting Mesozoic ruling reptiles have been popular with the public even since Henry Thomas De la Beche's 1830 giant illustration of them (see chapter 4). Here is a less congested and threatening Jurassic seascape. *Artwork by Elizabeth Virginia Stinchcomb.*

Glossary

Ammonoids. A subclass of extinct cephalopods having a (usually) coiled shell characterized by convoluted sutures. The ammonoids consist of the ammonites, the ceratites, and the goniatites (a Paleozoic group).

Ammonites. A class of cephalopods that have a shell characterized by complex sutures on the shell's chamber walls. Ammonites have the most complex shell sutures and are strictly Mesozoic fossils. Ammonites become extinct with the Mesozoic extinction event at the end of the Cretaceous period.

Archeological model (applied to collecting fossils). A restrictive model toward collecting which usually allows trained persons affiliated with an institution to engage in the extraction and collection of archeological materials. In archeology this is often necessary as a consequence of the limited and localized occurrence of artifacts at most archeological sites. Applied to fossils it is usually not a valid model as fossils usually occur in rock strata, which occur over a considerable area, the occurrence of fossils being where strata outcrop.

Biostratigraphy. The use of fossils for identifying or indexing rock strata. Biostratigraphy works because life through geologic time has changed so certain fossils can index or represent specific parts of geologic time and the strata deposited during this time.

Calcareous. Referring to the fact that a rock or sediment contains calcium carbonate, as in limestone and chalk.

Ceratites. A type (family) of coiled ammonoid cephalopod with a shell structure that is complex but lacks the convolutions characterized by an ammonite. Ceratites are characteristic of the Triassic Period.

Clastic (sediments). Sediments made of small particles such as clay, silt or sand grains; material derived from the weathering of previously existing earth materials. This is in contrast to chemical sediments like lime mud (the parent material of many limestone), which form from a chemical reaction.

Coprolites. Fossilized feces. Coprolites are sometimes capable of yielding information on what the maker ate. The most common coprolites are those of sharks; however, dinosaur coprolites are well known. Sometimes the coprolite is mineralized and attractive. Coprolites have been, at times, used as semi-precious gemstones. Agatized dinosaur coprolites are utilized by rockhounds in lapidary. Ear pendants made from ichyosaur coprolites containing iridescent ammonite shell fragments were popular in nineteenth century England.

Creationists. Individuals who hold the view that different forms of life (species) or that major categories of life (phyla and classes) were **independently** created by God or a supreme being. The opposite concept to creationism is transmutationism, the concept that life forms have transmutated or changed from one form to another.

Deeptime. A term referring to geologic time or megatime. In a space-time continuum paralleling deep space.

Druse or druze. A mass of small crystals that form in a cavity in rock. Specifically in reference to Mesozoic fossils the mass of small crystals that can fill the chambers of an ammonoid.

Evolute (ammonoids). A coiled molluscan shell in which the earlier formed portion of the shell (inner whorls) are exposed and are not covered by larger, later formed coils.

Facies. A specific type of sediment representing a recognizable environment in which a sediment (or sedimentary rock) was deposited.

Formation. A rock sequence that has some common, recognizable characteristic that can be recognized and which can be geologically mapped. Geological maps show the extent and age of geological formations.

Foraminifera. Small, single celled organism (eukaryotes) that has a calcareous test or "shell." Foraminifera have left an extensive and useful fossil record that is used extensively in biostratigraphy, particularly in the petroleum industry. Foraminifers (or forams), because of their (usually) small size, are generally not collectable fossils although they can be quite interesting.

Fossiliferous. Pertaining to sedimentary rock containing numerous fossils or composed of fossils, usually shell or shell fragments.

Goniatites. A group (family) of Paleozoic amminoid cephalopods with minimal complexity of the chamber-shell sutures. Goniatites range from the Devonian to the Permian.

Groundwater. Water saturating rocks which usually comes from surface precipitation. "Water in the ground."

Heteromorph ammonites. An ammonite having a shell that is configured in a manner other than the normal ammonite coil. Heteromorphs take on a variety of forms, one particularly strange form is where the shell curves back upon itself. Heteromorph ammonites are relatively uncommon and are generally highly collectable.

Involute ammonite (or amminoid). An ammonoid in which the inner (or earlier formed whorls) are covered by the outer whorls. In contrast to evolute amminoids.

Laterite (or lateritic soil). Soil or regolith of the tropics or a tropical environment. These are usually red in color and high in iron and aluminum. They are almost devoid of organic matter such as humus.

Marl. A soft, chalky clay of mudstone deposited under marine conditions. Marls can locally be rich in fossils such as oysters or other mollusks.

Megatime. Time spans in geology (and astronomy) which are measured in millions, tens or hundreds of millions of years and even billions of years. Such time spans are abstract and not part of a person's experience.

Oolites or oolitic limestone. Oolites are small spherical bodies usually composed of calcium carbonate and indicative of a shallow, warm-water marine environment.

Outcrop. A place where rock comes to the earth's surface, where it is otherwise usually covered by soil. Outcrops can be in creeks or other small streams, along rivers or in road cuts or quarries, anywhere rock is exposed at the earth's surface.

Paleontological Society. The scientific Society in North America dedicated to the study of a broad range of fossils and paleontology: both vertebrates and invertebrates as well as plants, monerans, and protists.

Phanerozoic. That part of geologic time when life was capable of leaving an obvious fossil record. The Phanerozoic consists of the Paleozoic, Mesozoic, and Cenozoic Eras of geologic time. Older earth rocks are Pre-Phanerozoic in age (or Precambrian in age after the earliest period of the Paleozoic Era, the Cambrian).

Placodonts. A group of peculiar, mollusk eating reptiles whose fossil remains are found in Triassic marine limestone, usually the Muschelkalk of central Europe.

Pre-Adamite "world." The geologic past from a creationist perspective. In the early nineteenth century two different views on the progression of life existed, that of the transmutationist and that of the creationist. The transmutationist believed in some sort of evolutionary transition between different forms of life. The creationists believed that God created major groups of organisms such as reptiles and mammals independently. The organisms which inhabited the world **before Adam and Eve** belonged to these previous creations, which in the early nineteenth century were being documented as fossils.

Radiometric age dating. The method of determining relative precise ages of rocks through measurements of their radioactive elements and the amounts of decay products. Radiometric age dates are generally considered the most accurate and reliable of various methods of dating geologic time and geologic phenomena.

Ruling reptiles. The large reptiles of the Mesozoic Era such as dinosaurs, plesiosaurs, and ichyosaurs, as well as less well-known forms such as the thecodonts and placodonts.

SAFE. Save America's Fossils for Everyone. A political action group headquartered in Cafifornia and concerned with amateur paleontological collecting, especially with regard to the collecting of fossil vertebrates.

Saurians. Usually large reptiles of the Mesozoic Era, identified by early "scientific creationists" – geologists of the early and mid-nineteenth century – as the highest forms of life at that time created by God. **Saurians** were believed to have been reptiles of a type higher than any of those of later geologic time or today, such as snakes, lizards, and turtles. These were believed to be degenerate forms from the once existing higher, but now extinct saurians such as dinosaurs, Mosasaurs, and ichyosaurs.

Silicified. Specifically a fossil that has been replaced with silica (quartz).

Stratigraphy (or sequence stratigraphy). The study of strata and enclosed fossils that occur in such strata as well as the age relationships of sedimentary rocks. The fundamental concept in stratigraphy is that when something is piled up, be it sediments or unanswered letters on a desk, those items at the bottom of the pile were placed there at an earlier time than those near the top.

Subcrop. Rock strata that are not exposed at the earth's surface but rather must be encountered by drilling or mining.

Sutures. The pattern made by the junction of a cephalopod chamber and the shell itself. Sutures are most complex, with many convolutions, in ammonites.

SVP. Society for Vertebrate Paleontology. A scientific society that is exclusively devoted to vertebrate fossils and vertebrate paleontology.

Synonymy A listing and dedication of scientific names (Linnaean names) which are believed by a competent paleontological worker to be invalid for a number of reasons, including the principle of publishing priority of a fossil's name.

Taxa. A unit of classification in organismic biology such as a species or other distinct category of life.

Taxonomy. The science (and art) of the classification of living things, which includes fossils.

Tectonism (Tectonic activity). Earth movements, which includes earthquakes and which are responsible for geologic forces such as uplift and mountain building.

Tethy's ocean. A large ocean existing during the Paleozoic Era and parts of which existed through the Mesozoic Era. Most of the sediments on the floor of this ocean have been destroyed by subduction, however some have been incorporated onto the edge of continents and the Tethy's ocean also spread onto continental shelves where these sediments (and their fossils) make up horizontal layers of sedimentary rock.

Thecodonts. An order of Triassic reptiles believed to be ancestral to the dinosaurs and birds. Thecodonts went extinct at the end of the Triassic Period.

Tracks. Trace fossils made by tetrapods of various types, such as dinosaurs, etc., which can be made on the surface of sediment, the surface then buried and the sediment lithified into rock. Triassic (and Jurassic) dinosaur tracks are probably some of the best-known fossil tracks and trackways.

Trackways. A series of tracks made by a tetrapod in the process of walking or running.

Transmutation. The concept that different forms of life (species) have transformed from one form to another by some process. Two forms of transmutation are generally recognized, Lamarckism (the inheritance of acquired characteristics) and Darwinism (the selection and survival of life forms through {usually} small changes known as mutations by natural selection, a type of natural selective breeding). The opposing concept to transmutation is creation by a supreme being.

Treatise of Invertebrate Paleontology. A series of definitive publications attempting to list all published genera of invertebrate fossils known up to the time of publication.

Tuffaceous. A sedimentary rock or sediment containing volcanic ash and usually deposited during a nearby volcanic explosion.

Undescribed fossil. A fossil that has not been recognized in the formal literature of paleontology.

Vertebrate fossils. Fossils (usually bones and teeth) of representatives of the Phylum Chordata.

Young earth creationists. Biblical fundamentalists who believe in a literal six days of creation as mentioned in the opening parts of the book of Genesis of the Christian Bible. Young earth creationists consider the earth to be very young, between 6,000 -10,000 years old.